Adaptive Learning by Genetic Algorithms

Analytical Results and Applications to Economic Models

Second, Revised and Enlarged Edition

Springer
Berlin
Heidelberg
New York
Barcelona
Hong Kong
London
Milan
Paris
Singapore
Tokyo

Herbert Dawid

Adaptive Learning by Genetic Algorithms

Analytical Results and Applications to Economic Models

Second, Revised and Enlarged Edition

With 52 Figures and 2 Tables

Springer

Dr. Herbert Dawid
University of Vienna
Department of Management Science
Brünnerstraße 72
A-1210 Vienna
Austria

ISBN 3-540-65607-3 Springer-Verlag Berlin Heidelberg New York
ISBN 3-540-61513-X 1st. ed. Springer-Verlag Berlin Heidelberg New York

Library of Congress Cataloging-in-Publication Data
Die Deutsche Bibliothek – CIP-Einheitsaufnahme
Dawid, Herbert: Adaptive learning by genetic algorithms: analytical results and applications to economic models; with 2 tables / Herbert Dawid. – 2., rev. and enl. ed. – Berlin; Heidelberg; New York; Barcelona; Hong Kong; London; Milan; Paris; Singapore; Tokyo: Springer, 1999
ISBN 3-540-65607-3

Printed in Germany

Hardcover-Design: Erich Kirchner, Heidelberg

SPIN 10715071 42/2202-5 4 3 2 1 0 – Printed on acid-free paper

To Laura

Preface to the Second Edition

The fact that I have the opportunity to present a second edition of this monograph is an indicator for the growing size of the community concerned with agent-based computational economics. The rapid developments in this field make it very difficult to keep a volume like this, which is partly devoted to surveying the literature, up to date. I have done my best to incorporate the relevant new developments in this revised edition but it is in the nature of such a work that the selection of material covered is biased by the authors personal interest and his informational constraints. My apologies go to all researchers in this field whose work is not or not adequately represented in this book. Besides the correction of some errors and typos several additions have been made. In the literature survey sections 2.4 (which was also reorganized) and 3.5 new material was added. I have also added a new section in chapter 3 which deals with the question how well empirically observed phenomena can be explained by GA simulations. A new section in chapter 6 presents a rather extensive analysis of the behavior of a two population GA in the framework of a sealed bid double auction market. Further minor additions and changes were made throughout the text.

I would like to thank all friends and colleagues who commented on the first edition of the book. Special thanks go to T. Fent, M. Kopel, A. Siebenförcher and T. Vallee for providing extensive error lists.

Vienna, October 1998

Preface to the First Edition

I started to deal with genetic algorithms in 1993 when I was working on a project on learning and rational behavior in economic systems. Initially I carried out simulations in an overlapping generations model but soon got dissatisfied with the complete lack of theoretical foundation for the observed behavior. Thus, I started to work on a mathematical representation of the behavior of a simple genetic algorithm in the special setup of an interacting population of economic agents and step by step arrived at the results collected here. However, I believe that much more can and has to be done in this field.

I would like to thank Gustav Feichtinger who not only supervised my doctoral thesis but always supported and encouraged me throughout the last few years. Special thanks are also due to K. Hornik, A. Mehlmann and M. Kopel who contributed largely to the work. During the preparation of the monograph I also benefited from helpful comments of A. Geyer-Schulz, G. Rote, G. Tragler and A. Rahman. Special thanks to W. A. Müller from Springer-Verlag for his support. Financial support from the Austrian Science Foundation under contract number P9112-SOZ is gratefully acknowledged.

Vienna, May 1996

Contents

1. Introduction 1

2. Bounded Rationality and Artificial Intelligence 7
2.1 Bounded Rationality in Economics 7
2.2 Artificially Intelligent Agents in Economic Systems 11
2.3 Learning Techniques of Artificially Intelligent Agents 13
2.3.1 Genetic Algorithms and Related Techniques 13
2.3.2 Classifier Systems 14
2.3.3 Neural Networks 17
2.3.4 Cellular Automata 21
2.4 Some Applications of CI Methods in Economic Systems 26
2.4.1 Bidding Strategies in Auctions 26
2.4.2 The Iterated Prisoner's Dilemma 27
2.4.3 Goods Markets 31
2.4.4 Financial Markets 33
2.4.5 Evolving Market Structures 34
2.4.6 Further Simple Economic Models 36
2.5 Potentiality and Problems of CI Techniques in Economics 37

3. Genetic Algorithms 41
3.1 What are Genetic Algorithms? 41
3.2 The Structure of Genetic Algorithms 42
3.3 Genetic Operators 43
3.3.1 Selection 43
3.3.2 Crossover 45
3.3.3 Mutation 46
3.3.4 Other Operators 46
3.3.5 An Example 47
3.4 Genetic Algorithms with a Non-Standard Structure 48
3.5 Some Analytical Approaches to Model Genetic Algorithms 49
3.5.1 The Schema Theorem 50
3.5.2 The Quantitative Genetics Approach 52
3.5.3 Markov Chain Models 55
3.6 Do Genetic Algorithms Describe Adaptive Learning? 61

3.6.1 Genetic Algorithms and Empirical Evidence 61
3.6.2 Economic Interpretation of the Algorithm 65

4. Genetic Algorithms with a State Dependent Fitness Function 71
4.1 State Dependency in Economic Systems 71
4.2 A Markov Model for Systems with a State Dependent Fitness Function 72
4.3 The Difference Equations Describing the GA 77
4.4 Deviation from the Markov Process 78
4.5 A Numerical Example 82
4.6 Stability of the Uniform States 83
4.7 Two-Population Models 92

5. Genetic Learning in Evolutionary Games 97
5.1 Equilibria and Evolutionary Stability 97
5.2 Learning in Evolutionary Games 98
5.3 Learning by a Simple Genetic Algorithm 101
5.3.1 Rock-Scissors-Paper Games 103
5.3.2 A GA Deceptive Game 107
5.3.3 Learning in Non Deceptive Games 111
5.4 Two-Population Contests 114

6. Simulations with Genetic Algorithms in Economic Systems 121
6.1 A Model of a Competitive Market 121
6.1.1 Pure Quantity Decision 124
6.1.2 Exit and Entry Decisions 127
6.2 An Overlapping Generations Model with Fiat Money 132
6.2.1 Learning of Cyclical Equilibria 136
6.2.2 Learning of Sunspot Equilibria 140
6.3 A Sealed Bid Double Auction Market 143
6.3.1 Constant Cost and Utility Values 143
6.3.2 Stochastic Cost and Utility Values 155

7. Stability and Encoding 161
7.1 The Cobweb Example Revisited 161
7.2 Impact of a Change in Encoding and Scaling 165
7.3 A Method for Finding Economic Equilibria 167

8. Conclusions 169

A. Basic Definitions and Results Used 171
A.1 Time Homogeneous Markov Chains 171
A.2 Nonlinear Difference Equations and Stability 173

B. Calculation of the Equilibria of the Evolutionary Games in Chapter 5 175
B.1 Rock-Scissor-Paper Games 175
B.2 The GA Deceptive Game GAD 177
B.3 The Games G1 and G2 178

C. Proof of Proposition 6.3.1 181

References 183

List of Figures 193

List of Tables 197

Index 199

1. Introduction

The analysis of mathematical models describing the learning behavior of rational agents has been one of the major topics of economic research for the last few years or even decades. Many different models have been proposed and their analysis has given the researchers some insights into the phenomenon of formation of equilibria in economic systems. On the other hand, the modern development of computer technology has caused the rise of a new dynamic field of research, which deals entirely with the understanding and imitation of human behavior, namely the "artificial intelligence" research. Although there is considerable overlap between these two fields they have long developed in isolation from one another. Only recently the interest of some economists in certain techniques mainly from the AI related field of "computational intelligence" (CI) has increased and has led to the applications of these techniques to economic models.

The main reason for the weak interaction between economists and CI researchers is that the tools and goals of these two groups are quite different. The traditional approach of economists is to analyze economic systems with the help of mathematical theory. Relying on plausible behavioral assumptions about economic agents an adaptive learning model is constructed. The economists use mathematical representations of the model and try to derive analytical results for these models. In order to keep these models analytically tractable most of these models use rather simple behavioral assumptions. Nevertheless, a large number of conclusions could be derived with this approach. Often the mathematical analysis allows structural insights and explains similarities and differences in the behavior of different models. On the other hand, the majority of models allow only a few weak or local results. In these cases a mathematical analysis draws only a coarse picture of the learning behavior.

Quite the opposite holds true for the approach taken by computer scientists. They like to deal with learning models which may be efficiently implemented on a computer. Usually mathematical considerations are of minor importance, and the algorithms rely on heuristic arguments and similarities to nature. Normally the analysis of these algorithms is done by testing them with a large number of real world or benchmark problems. The obtained numerical results are used to build up conjectures regarding the performance of the algorithm in different set-ups. On one hand this approach allows to use more complex learning models than the analytical approach described above but on the other hand simulations can never *prove* a certain feature of the model but only suggest it. In general the structural insights which may be obtained by numerical simulations are not comparable to general mathematical results. However, simulations allow some ideas about the behavior even if a mathematical analysis is completely impossible.

In this monograph I connect these two approaches and analyze a special learning algorithm developed in the CI literature with analytical methods and simulations. I concentrate on genetic algorithms (GAs) and show how these learning algorithms may be interpreted as models of the learning behavior of a population of adaptive agents. Further, I translate this algorithm into a framework of traditional learning models in economics, namely mathematics. Doing so allows to take both the analytical and numerical approach in order to study this algorithm. In the analytical part a more or less standard analysis of an economic learning rule is carried out and the long run behavior of a population which acts according to this learning model is described. Markov theory is used to derive a result which characterizes the possible long run states of the population. In analyzing a system of difference equations, I furthermore obtain conditions for local stability or instability of certain population states. Based on these results it is easy to realize the importance of the calibration of processes like genetic algorithms. Different coding mechanisms and parameter values may lead to a completely different long run behavior of the system. These results will be first shown by mathematical proofs and also illustrated by examples from the fields of game theory and economics.

However, the part of these notes where simulations are presented is not intended as a pure illustration of the mathematical results, but should be regarded as an important part of the analysis of GAs in economic systems. The simulations show several properties of the process which could not be proven analytically and I also present simulations in models where the assumptions needed for analytical results are not fulfilled. Thus, in these notes both the analytical approach and the simulation approach are of great importance. Both approaches together give quite a good picture of the behavior of GAs in economic systems.

Besides the technical aspects a major goal of these notes is also to give an economic interpretation of the GA as a learning rule. Some words of motivation are in order here. Why do we consider GAs as a model of adaptive learning and why should this analysis be of interest. First of all, as already mentioned and as elaborated in chapter 2 various researchers have used GAs for simulating the behavior of a population of interacting agents. Although this per se should not be seen as a striking argument that GAs are a useful model of learning the analysis carried out here can in any case be seen as a theoretical foundation for these results. However, it is maybe even more important to gain some understanding which behavioral assumptions underlie the model. If we know how to interpret the different parts and parameters in the GA from an economic point of view, theoretical results concerning the behavior of the process may be used to gain additional insights into the relation between behavioral assumptions in the model and features of the learning process. When speaking about features of the learning process I am referring to mathematical features of the dynamical system like local stability or instability of economic equilibria or non-equilibrium states or global features like complex or chaotic behavior. But exactly these relations allow us to make general statements about the structure of learning processes. Thus it must be a major goal of evolutionary economics as such to look for such general relations.

Although it will become obvious that Genetic Algorithms are motivated rather by biological than by economic considerations it captures essential features of adaptation which are also highly relevant in an economic context. A task of this monograph is to point out the economic interpretation of the algorithm and use these insights to derive an economic message from the analytical and numerical results. The main goal of such a study should be the better understanding of adaptation and I am confident that the results presented here indeed allow some basic insights into the learning behavior of adaptive and bounded rational individuals.

On the other hand, Genetic Learning in economic systems may also be seen from a completely different point of view. In systems with a very complex structure it may be impossible to calculate the economic equilibria of the system analytically. In such a situation it may be a promising approach to use a numerical simulation of a learning population and observe the behavior of such a system. If the adaptation process converges towards some fixed point this point is clearly a candidate for an equilibrium. It might be of some interest that an argument like this lead to the introduction of the concept of fictitious play (see chapter 5) into evolutionary game theory. Although the process has a clear economic interpretation it was first introduced as an algorithm to solve games. In chapter 2 we will see that also GAs have already been used to determine previously unknown equilibria in economic systems. However we will see that – contrary to fictitious play – not every limit point of a GA is necessarily an economic equilibrium. Thus we have to have some criterion to determine whether a fixed point is an equilibrium or not and, if it is not, we have to know how to facilitate the convergence of the GA towards an equilibrium. Such a criterion will be inferred and provide some guidelines for shaping the GA in a way such that the algorithm has desirable dynamic properties.

Finally, I like to point out at this stage that there is a basic difference between the analysis carried out in this book and the existing results about the behavior of GAs. We will see in chapter 4 that the fitness value of a string depends in general on the state of the population if the GA is used for a simulation of an economic system, whereas the fitness value is independent of the state of the population if the GAs is used for optimization. This basic difference makes it impossible to apply existing theoretical results to the systems studied here. To my knowledge this book is the first one where this difference is explicitly emphasized and where theoretical results are derived for the more complex situation where the fitness of a string depends on the state of the whole population. However, the findings presented in this book are of course also valid for the traditional field of application of GAs, namely optimization problems. Thus, this work may be of interest as well for economists as for mathematicians and computer scientists.

The monograph is organized as follows. In chapter 2 I discuss the importance of adaptive learning models in economics and give a short overview of the existing literature dealing with the application of CI techniques in economic systems. Although it is not the purpose of this monograph to give an introduction to CI techniques I give a very short description of the most important techniques before discussing some of their applications in economics. This short introduction should allow readers who are not familiar with CI concepts to grasp their basic underlying ideas, and, on the other hand, should give readers with a computer science background an idea of the questions which are relevant from an economic point of view. In chapter 3 an introduction to genetic algorithms is given, the most important analytical results concerning GAs are reviewed and the relevance of this learning model for economics is discussed. This introduction will be more detailed than the general introduction in chapter 2. People familiar with GAs could skip this chapter except subsection 3.5.3. In this subsection the basic model and the notation used throughout the text is presented. Chapter 4 is the most important part of the monograph. In this chapter all analytical results describing the behavior of genetic algorithms in economic systems are derived. These results are illustrated by various simulations in the subsequent two chapters. Chapter 5 shows the learning behavior of GAs in evolutionary games. The simulation results are compared with stability concepts from evolutionary game theory and the effects of different parameter constellations are investigated. Chapter 6 provides three examples of genetic learning in standard models from microeconomics. First I analyze a model of a competitive market and show in various simulations the importance of the theoretical results from chapter 4. As a second example I present simulations done in an overlapping generations framework with cyclical and also stochastic equilibria. The second example shows that genetic algorithms are well suited to learn also non-trivial equilibria in rather complex systems. The third part of this chapter contains an analysis of genetic learning in a double auction market. Here two populations interact and we have the opportunity to explore the potential of theoretic analysis in such a framework. An exposition of how the results from chapter 4 may be used to facilitate the learning of economic equilibria by genetic algorithms is given in chapter 7. Finally, an outlook to the possible further development of this line of research and its importance for economics constitutes chapter 8. In the appendices I provide the most important definitions and results from the theory of Markov chains and difference equations which are used in the text.

2. Bounded Rationality and Artificial Intelligence

2.1 Bounded Rationality in Economics

The traditional and most widely used approach for the analysis of economic systems is concentrated on equilibrium behavior. We may say that an equilibrium in the broadest sense is a situation where no agent has any incentive to deviate unilaterally from the current behavior. There are several equilibrium concepts for different classes of economic models but all these concepts rely on similar assumptions about the rationality of the economic agents. Basically, two assumptions have to be made in order to state that an economic system will a priori be in equilibrium. First we have to assume that all agents are willing and able to maximize their expected utility and second that all agents have rational expectations. Rational expectations means that all agents have identical and exactly correct beliefs about how everyone will behave. Agents who fulfill both assumptions are often called completely rational.

To illustrate these assumptions let us consider a duopoly of firms. Both firms produce the same good and the price of the good is determined by the aggregate supply of the good. Suppose the two firms have no opportunity to make a binding agreement about their quantity of output. Thus, none of the two firms knows the output of the other firm and therefore also the price of the good in the next period is unknown to the firms. Nevertheless, the firms have an expectation about the other firms' output and the price of the good in the next period. The assumption of perfect rationality would imply that both firms are able to anticipate the correct output value of the other firm. The assumption of complete rationality implies further that they have to be able to determine their own optimal output value given this expectation about the other firms output. However, in order to calculate the optimal output value the firm has to know the effects of the aggregate supply on the price. In other words, the firm has to have exactly correct beliefs and has to know the structure of the system, in particular the price-demand relationship of the good. Of course it could be argued that the demand curve can be inferred from previous data; but, on one hand, this can only be an econometric estimate of the correct function, and on the other hand, the demand curve might shift in the course of time. Even for a monopolistic firm it is by no means trivial to choose the optimal output policy if it does not know the exact demand curve and certainly things get much more complicated if more than one firm is involved.

As can be seen in this example the assumption of a completely rational firm is quite demanding and not very realistic. Nevertheless, economists have based most of their considerations on the assumption that an economic system is always in equilibrium. Much of their work was devoted to the question of how the equilibrium behavior changes when parameters in the model change – this kind of analysis is called comparative static – but only very little to the problem how the transition from one equilibrium to the other actually takes place. Obviously, a change of model parameters pushes the system out of equilibrium even if it was in such a state before. Without the assumption of complete rationality it is not clear why the system should immediately approach the new equilibrium as is stated in the comparative static analysis. Of course, there are some traditional justifications for the comparative static approach like the Walrasian tâtonnement process which takes place in virtual time, but basically any analysis which ignores out of equilibrium behavior uses the implicit assumption of perfect foresight and complete rationality. However, in the recent years the effort spent for understanding out of equilibrium behavior has grown a great deal.

Whereas the assumption of maximizing behavior of economic agents is quite unchallenged in the literature the discussion of the assumption of rational expectations on the agents' side was very active. One explanation for this phenomenon may be that in many setups almost any policy or rule of thumb used for decision making can be interpreted as the optimal behavior under some expectations (Bernheim [22], Pearce [136]). Thus, from a formal point of view it is in many cases sufficient to discuss the effects of the relaxation of the assumption of rational expectations. Several scholars dealing with economic or game theoretic models have pointed out that the assumption of rational expectations assumes too much sophistication and coordination of beliefs on the part of decision-makers. Although a similar point of view is adopted in this book, I would nevertheless like to point out that there are several arguments which may be used to motivate the assumption of rational expectations.

The first argument often used in game theoretic models is the story of preplay communication. Suppose that all agents meet before they decide on their actions and talk about their future steps. If they agree on some behavior, this set of actions must be an equilibrium (otherwise at least one agent has an incentive to deviate) and it is completely rational to have strong confidence that these actions will actually be taken. However, somehow this story shifts the problem just one level up. If there exist several equilibria where none of them is Pareto dominant it is by no means clear how preplay communication should lead to an agreement. Thus, in this case the preplay communication stage has to be included into the model and does not justify the assumption of rational expectations.

Another argument in favour of the use of rational expectations is that there are situations where all agents a priori know the outcome of the game. Consider a situation where all the agents have the skills to calculate the equilibrium and that this fact is common knowledge (e.g. all agents have read the same economic or game theoretic text book). In such a situation everyone may expect that all others behave according to this equilibrium and follow the equilibrium himself in order to react optimally to this expected behavior of the others. Thus, rational expectations may occur in equilibrium if all agents are able to determine the equilibrium and know that all the others are able to do so. However, the assumptions which are implicitly made in this story are again quite strong. First of all, there must exist a unique equilibrium of the system. If this holds it is still not clear how the players get the necessary information in order to determine the equilibrium and why they know that all the other agents have the necessary information and skills too.

Of course these are not the only two possible justifications of rational expectations but they show that a justification of rational expectations is always founded on very strong assumptions about the knowledge and skills of the different agents. On the other hand, the concept of rational expectations and equilibrium models in general have been used with great success in a large number of economic problems. Thus another justification of the assumption of rational expectations – and obviously a very good one – is the simple fact that it works well in describing several aspects of the evolution of an economy. Putting together these arguments we may conclude that rational expectations are a very useful assumption in order to solve models and get certain predictions about basic economic relations. However, the used assumptions are quite unrealistic in real world applications. Or, as Lucas [111] puts it *" We use economic theory to calculate how certain variations in the situation are predicted to affect behavior, but these calculations obviously do not reflect or usefully model the adaptive process by which subjects have themselves arrived at the decision rules they use. Technically, I think of economics as studying decision rules that are steady states of some adaptive process, decision rules that are found to work over a range of situations and hence are no longer revised appreciably as more experience accumulates."*

Arguments like the one of Lucas cited above have brought many economists to think about the actual adaptation processes which appear out of equilibrium. By relaxing the assumption of rational expectations these models try to explain how equilibrium behavior can emerge as the limit of an adaptation or learning process of bounded rational agents. Obviously, several questions arise. First of all it is by no means clear in which ways the assumption of correct beliefs should be relaxed. We could think of adaptive agents who build their expectations about the actions of the others according to Bayesian learning and act optimally with respect to these expectations or we could think of agents who just adopt the last action of some other agent. These two scenarios use quite different assumptions about the agents rationality and will of course lead to quite different results. In order to formulate a model of adaptive learning, the following two questions have to be addressed. How do the agents update their expectation about the behavior of their surrounding and how do they react to their expectations? In particular, are they able to act optimally under the assumption that their expectations are right? If we have formulated a model of adaptive behavior the key question is: What happens in the long run if all agents indeed learn according to this rule? In particular, does this kind of adaptation lead to equilibrium behavior or do at least the beliefs of the agents converge towards an equilibrium? Convergence results of this kind are very important for economic theory as they show that equilibrium behavior and rational expectations might emerge from adaptive behavior of bounded rational individuals. This would provide another justification for the use of the assumption of rational expectations.

The previous paragraphs show quite clearly that there are many ways to describe the adaptive behavior of bounded rational agents and this fact also becomes apparent if we look at the large number of different models analyzed in the literature. Some of these models assume that the agents in the model behave just like econometricians and build up their beliefs according to econometric estimations that are based on observed data. Examples for models with this background are least square learning (e.g. Sims [160], Marcet and Sargent [114], [115]), stochastic approximation (Robbins and Monro [141], Woodford [186]), Bayesian learning (e.g. Blume and Easley [28], Turnovsky [167]) or simple moving average updating (Lucas [111]). Additionally, there is a very active branch of game theory, called evolutionary game theory, which deals with learning in iterated games (see Chapter 5). Of great importance for the theory of bounded rationality is also the approach of Simon ([159]). He claims that individuals are "satisficing" their needs, which means that they are content with good but not optimal actions. If they are content they stop looking for better solutions, otherwise they adopt some other action with the help of some not specified algorithm.

This plethora of models is a severe drawback of the whole project to describe boundedly rational behavior. Our understanding of the observed adaptive behavior is obviously not good enough to create one valid reference model of adaptive behavior in economic systems. Unfortunately, there is only very little solid empirical foundation for the different hypotheses about how expectations are formed. This lack of empirical work has been pointed out by several scholars, e.g. Simon [159] states that "*Armchair speculation about expectations, rational or other, is not a satisfactory substitute for factual knowledge as to how human beings go about anticipating the future, what factors they take into account, and how these factors, rather than others, come within the range of their attention.*" As long as no rigorous empirical foundation of the forming of expectations exists we have to analyze models which describe some plausible features of actual learning behavior and show whether such models are suited to motivate the equilibrium concepts used.

The problems which may occur in such an ad hoc approach will become quite apparent in the analysis carried out in this book. Nevertheless, I believe that at this stage such an approach may allow new insights into the dynamics of actual adaptive behavior.

To finish this section I would like to point out several publications where the topic of bounded rationality is discussed in more detail. Sargent [150] gives a very good introduction into the topic of bounded rationality and deals extensively with artificially intelligent agents. Some of the ideas of Simon regarding this question are lined out in Simon [159] where a large number of his articles relevant for economics are collected. A recent book by Kirman and Salmon [97] contains a collection of contributions of several leading scholars in the field of learning and adaptation in economics. Two not so recent articles, which are however in my opinion still of great interest, are by Day [52] and Lucas [111]. The interested reader should also look into the special issue 'On the Economics of Limited Cognition' of the *Journal of Economic Behavior and Organization* (Vol. 29, No. 2, 1996). From a large number of publications dealing with adaptation and learning in game theoretic models I would like to point out some survey articles like the ones by Selten [156], Mailath [112] or Nyarko et al. [135] but also a very nice book of Binmore [29] and recent ones by Fudenberg and Levine [62], Samuelson [147], Vega-Redondo [172] and Weibull [180].

2.2 Artificially Intelligent Agents in Economic Systems

In the last section I have tried to sketch the argument which led to the development of the theory of adaptation and learning in economic systems. However in this monograph I will not deal with traditional economic learning models but with artificially intelligent agents. So in this section I will point out the basic properties of artificially intelligent agents and figure out the most important differences between this approach and the standard economic learning models.

Concentrating again on the class of models discussed in the last section we realize that in many of these models a whole population of agents is represented by one representative individual. This representative individual updates his beliefs according to the chosen learning rule and determines the state of the system in the next period by the action taken. The main advantage of this approach lies in the fact, that the resulting systems can be analyzed analytically and that a number of convergence and stability results have already been derived for systems of this type. On the other hand, it is highly stylistic to model a population of completely identical individuals, who all base their actions on identical anticipations of the same parameters. If the expectations of the representative individual is interpreted as the average expectation of all individuals the best reply to this average expectation does in general not equal the average of the best replies to the individual expectations. This equality only holds true if the reaction function is linear in the expectations. Thus, in general an important effect is neglected if only representative individuals are considered instead of whole populations. Furthermore, the use of a representative individual does only allow learning by reaction to observed variables of the system like prices or population strategies. However, it is impossible in this setup to model individual interaction between agents.

Motivated by these and similar arguments the development of modern computer technology has established a new line of research in modern economic theory, namely the introduction of artificially intelligent agents to economic systems[1]. Although the field of artificial intelligence has by no means reached the goal to build entire artificial beings comparable to humans, the techniques developed in this area are well suited to imitate human learning in simple models.

Let us first clarify what we mean by artificially intelligent agents. An artificially intelligent agent has a much higher computational capability than the traditional economic agent. Whereas the economic agent reacts to signals from the economic system in a functional way exactly determined by the model, an artificially intelligent agent has some interior mechanism determining his reaction to incoming signals. The way the artificially intelligent agent reacts to the same signal changes according to the data previously observed. Thus, learning in this setup means that the decision rule per se is changed in the course of time. Contrary to this most of the standard economic learning rules have a fixed decision rule and update only the expectations about the actions of the competitors. If we consider for example a simple learning rule like moving average learning in a market situation, we typically have a fixed reaction function according to which the firms react to price expectations. Only the expectations are updated according to the moving average rule. We will see that the various techniques used to create artificially intelligent agents differ quite strongly; however, the basic distinction from standard learning techniques in economics is that the decision rules of the individuals are not implemented a priori but evolve in the course of time.

The computational capacity of modern computers allows us to build whole populations of such artificially intelligent agents, and to watch their behavior in a given system. The most widely used techniques are neural networks (NN), genetic algorithms (GA), classifier systems (CS) based on genetic algorithms or cellular automata (CA). In the modern literature all these techniques are subsumed under the term "computational intelligence" (CI) methods to distinguish these techniques depending on numerical data from knowledge handling techniques like expert systems which are referred to as "artificial intelligence" techniques (see Bezdek [24], Marks [117]). All the same, we refer

[1] The community of researchers dealing with 'Agent Based Computational Economics (ACE)' is very vivid and fast growing. The interested reader should visit the ACE Web page at http://www.econ.iastate.edu/tesfatsi/ace.htm

to the individuals whose learning behavior is modeled by one of these techniques as artificially intelligent agents. This is also standard in the existing economic applications.

In the rest of this chapter I will first give a very basic description of the CI techniques cited above and present a survey of applications of these techniques in various economic models. Afterwards I will discuss the advantages and disadvantages of the use of artificially intelligent agents. I do not restrict my attention to genetic algorithms in this section because I believe that the basic properties of artificially intelligent agents are quite independent from the actual algorithm used.

2.3 Learning Techniques of Artificially Intelligent Agents

The short introduction to the different CI techniques given below is only intended as a help to enable readers not familiar with these algorithms to understand the results and discussions presented in the next sections. Of course any reader with a deeper interest in this topic should read some of the huge amount of introductory literature. Some references are given in the text.

2.3.1 Genetic Algorithms and Related Techniques

Since an introduction to GAs is given in chapter 3 only a few words are said here. A Genetic Algorithm is a search algorithm based on the principles of natural evolution. A population of strings is considered where each string encodes an admissible input of the surrounding system. This input may be a solution of an optimization problem or – as in economic systems – the decision (rule) of an economic agent. Every input gets some reward from the surrounding system, the so called fitness value of the string. Using these fitness values a new population of strings is generated by applying some genetic operators to the old population, see chapter 3 for details. Starting with a randomly initialized population this iteration is carried out for a given number of periods. Closely related algorithms are evolutionary strategies (ES) (Schwefel [153], Rechenberg [139]) which use slightly different operators than GAs and genetic programming (GP) (Koza [103]) which extends the genetic model of learning into the space of computer programs. In most applications the programming language Lisp is used. Finally, there is also evolutionary programming (EP) (Fogel and Atmar [61]). There are two main differences between GAs and EP. First, EP uses no crossover operator and second there is no constraint on the representation in EP. Unlike in GAs no linear representation has to be used since no crossover is used. Thus, this algorithm is also well suited to describe to evolution of structures like finite automata or neural networks.

2.3.2 Classifier Systems

Classifier Systems (CS) were first introduced by Holland [87]. The basic idea of this algorithm is to give an explicit model of the interaction between the signals that an individual receives from the environment, its decision making, and the signals sent to the environment. Typically, a CS consists of the following parts: an input interface (also called detectors) where the messages or signals from the environment enter the system, a decision unit where it is determined as how to react to the incoming messages, and an output interface (or effectors) which implements the chosen action. Of course, the interesting point about this setup is how the decision unit works.

Holland considers the following setup. Let us assume that the input interface translates the incoming signals into binary strings and passes these strings on to the decision unit. The decision unit consists of two parts the message list and the rule list . Whereas the message list contains messages coming from the input interface or the decision unit itself (all these messages are binary strings of a fixed length) the rule list consists of a given number of strings which are interpreted as rules. In the simplest setup a rule has one condition part and one action part. Each of these parts is a string of the same length as the messages, which implies that a rule is a string with double length of a message. Whereas the action part of the string is always binary the condition part is a string over the alphabet $\{0, 1, \#\}$ where # is the so called "don't care" symbol. The fewer "don't care" symbols are in the condition parts of a rule the more specific this rules is. If a message enters the decision unit, it is added to the message list. Afterwards, it is checked for all rules in the rule list whether their condition parts are matched by some message in the message list. Matching simply means that there is some binary string in the message list which equals all specified bits of the string acting as the condition part of the rule. The bits with "don't care" symbols are irrelevant as the don't care symbol is matched by 0 and 1. In many applications the rules have more than one condition part. In this case the rule can only be active if every condition part is matched by some message on the message list. Figure 2.1 shows a very simple case with four messages in the message list and 4 rules with two conditions in the rule list. The first and last of the rules are matched in this situation.

Next, it has to be determined which of the active rules should post the message in its action part. This decision is made randomly but not all rules are chosen with the same probability. Each rule has some strength and in general the probability to be chosen to implement its action is positively correlated with the strength and the specifity of the rule. As how the strength of a rule is determined will be explained just below. If a rule has been chosen to post its message, the message is added to the message list. There are rules where the message in the action part is an output message. In this case the message is sent to the output interface and triggers some action. On the other hand,

message list	rule list		
	cond1	cond2	action
0100	0#00	1000	0000
1000	#010	010#	0101
0101	0110	1111	1111
1111	1000	0101	0010

Fig. 2.1. A simple example of a message and a rule list in a classifier system; the first and the fourth rule are matched

there are also rules whose action part contains an internal message where no action is taken at the moment.

In many CS the strength of the rules is governed by the so called *Bucket Brigade Algorithm* . Whenever a rule posts its message the strength of this rule is updated. On one hand, a part of the strength of the rule has to be transferred to the rules whose action part matched the condition part of the rule, on the other hand, the rule earns new strength from the posted message. If the message is an output message then the increase of the strength of the rule depends on the payoff of the action triggered by this message. If the message is an internal message then this message is added to the message list and the rule gets new strength whenever this message matches the condition part of the rule which is chosen to post its message. Thus, the rules may be seen as middlemen in an economy who supply and consume messages for some price. The idea behind this algorithm is of course that rules which lead to actions with a high payoff have a high strength and are allowed to post their messages more often. The problem of the determination of the strength of the rules in general is often referred to as the credit assignment problem.

The system described here is an algorithm to create outputs in correspondence to inputs; however, the adaptive ability of this algorithm is very restricted up to now because the set of rules is fixed and the only adaptation to the environment is that the more useful rules accumulate a higher strength. To admit the detection of previously not existing rules, a new set of rules is created out of the old one after a certain number of periods. This updating of the set of rules is done by using a genetic algorithm. The strength of the rules is used as the fitness of the corresponding string. Using this algorithm the bad rules should vanish from the rule list and the genetic operators applied to the old rule list should create new useful rules. Now we have a complete classifier system. Summarizing, the basic execution cycle of a classifier system contains an iteration of the following steps:

1. Messages from the environment are placed on the message list.
2. Each condition of each classifier is checked against the message list to see if it is satisfied by (at least one) message thereon.

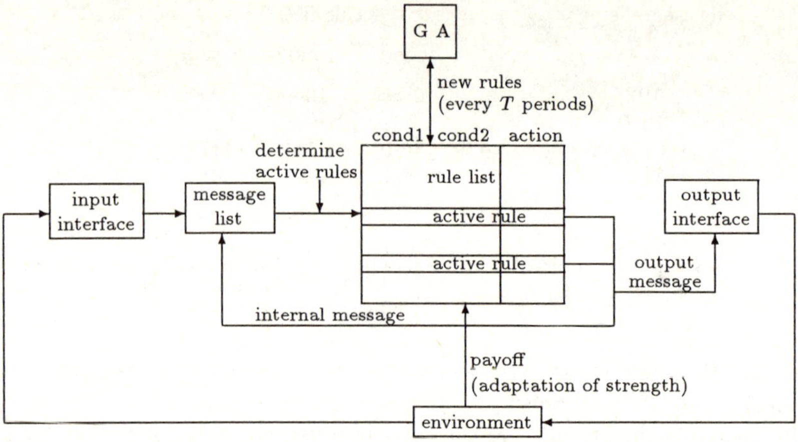

Fig. 2.2. The structure of a classifier system

3. All classifiers that have all conditions satisfied participate in a competition and those that win post their message to the message list.
4. All messages directed to effectors are executed (causing actions in the environment).
5. All messages on the message list from the previous cycle are erased (i.e. messages persist for only one single cycle, unless they are repeatedly posted).

In Figure 2.2 the overall structure of a CS is depicted. Consult Holland [88] for a more extensive introduction to classifier systems.

Obviously a CS models the interior decision making of one individual decision maker. So if we talk about artificially intelligent economic agents a CS represents exactly one artificially intelligent agent. By creating such an artificially intelligent agent we can only determine to which aspects of his environment he may respond, but a priori we do not know in which way he will make his decisions. Let us again consider the model of a duopoly market. It seems to be quite a natural approach to represent any of the two firms by one classifier system. Basically, we have to decide on two facts before we construct these two CS. First, what are the input messages that the CS should receive from the environment, and second, what are the output messages. The second question can be answered easily. Either we interpret the output messages as the quantity sold or as the price of the good. This is just the distinction between Cournot and Bertrand duopoly. Concerning the input we have several possibilities. We could just encode the current price of the good in the input message or we could encode just the amount of the good currently offered. Of course we could also encode both informations, price and amount of the good. Perhaps this would facilitate the estimation of the demand price function. Another possibility would be to encode the amount currently offered by the competitor, either alone or together with some other information. The more information we give to the system the more sophisticated behavior we may expect. On the other hand, a lot of information implies a long message string and the amount of time till good rules are found may increase dramatically. So as in any of these problems we have a trade-off between the quality of the solution and the effort invested. Note that the structure of the evolving reaction functions of the two agents is completely free a priori. It is also not clear a priori whether the same kind of reaction functions will evolve in different simulation runs or whether the system will approach the same stationary condition every time. Finally, I would like to point out that this kind of algorithm is well suited to model the interaction between a small number of interacting agents but that the modeling of a large population is not very convenient to implement. Any single agent has to be represented by a complete classifier system with a message and rule list. Thus, the demand for memory and computation time is large for large population sizes.

2.3.3 Neural Networks

Neural Networks (NNs) are maybe the most popular tool created by AI research. The basic idea of the development of neural networks was to mimic the way of working of the human brain. It was recognized that the single cells of the brain have a relatively slow rate of operation but that the parallel working of a huge number of highly interconnected cells makes the brain so efficient. Thus, parallelism is one major property of NNs. In fact NNs were used with large success for problems like pattern recognition, forecasting, signal processing or any mapping problem where lots of example data are available.

As already mentioned, the NN tries to imitate the human brain. Thus, the used terminology is influenced by biology. A neural network consists of several connected cells called neurons. A neuron contains three basic elements. First, there are the synapses or connecting links coming from the other neurons. Each link is characterized by its weight w_{kj} where neuron k is at the output end and neuron j is at the input end of the link. Every signal x_j sent by neuron j to k is multiplied by this weight and arrives at neuron k as signal $w_{kj}x_j$. Weights may be positive or negative. If the weight is positive the link is called excitatory if the weight is negative the link is called inhibitory. Second, all incoming messages are simply summed up. This sum is often called activation level. Third, there is a so called activation function which transforms the activation level to some output of the neuron. In general the outputs are restricted either to the interval $[0, 1]$ or to $[-1, 1]$. Often some threshold value θ_k is subtracted from the sum of inputs before the activation function is applied. Thus, the input output correspondence of a neuron may be written mathematically as

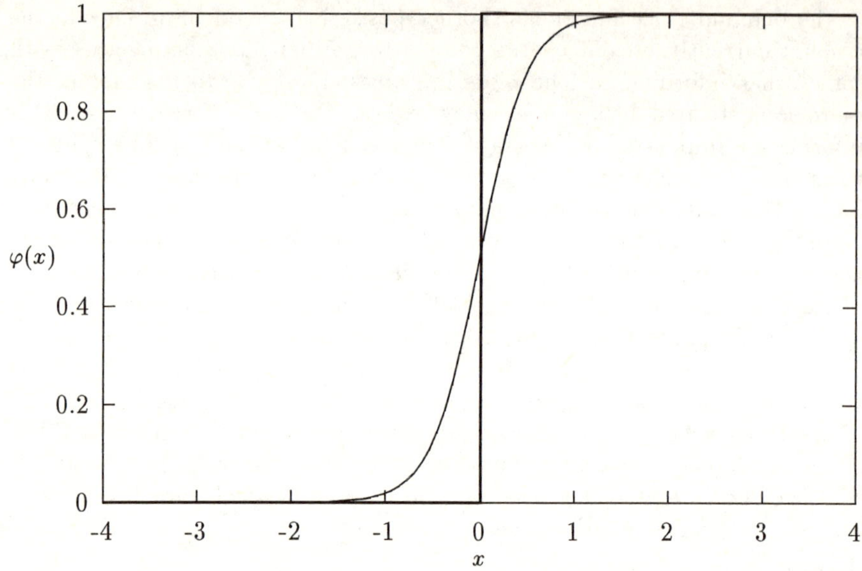

Fig. 2.3. Example of a threshold and a sigmoid logistic activation function

$$y_k = \varphi\left(\sum_{j=1}^{p} w_{kj}x_j - \theta_k\right),$$

where $x_j,\ j = 1, \dots, p$ are the outputs of all neurons connected with neuron k, y_k is the output of neuron k and $\varphi(\cdot)$ is the activation function.

There are several types of activation functions. The most common are threshold functions, also called heaviside functions, or sigmoid activation functions. A sigmoid function is a strictly increasing smooth function approaching asymptotically the values 0 and 1 if the argument tends towards plus or minus infinity respectively. From a mathematical point of view dealing with a sigmoid activation function may be much more convenient as this function is, contrary to the heaviside function, continuously differentiable. Examples of both types of functions are given in figure 2.3. Note that a threshold function can assume only 0 or 1 as values whereas the image of the sigmoid function is the whole interval $[0, 1]$.

The architecture of a NN is determined by the way the different neurons in the net are interconnected. Basically, we have to differentiate between two types of networks, namely feedforward networks and recurrent networks. In a feedforward network no cycles of links exist. There are input nodes which absorb a signal from the environment. Afterwards, this signal is transferred and propagated until it has reached an output node which produces an output to the environment. No signal emitted by a neuron can come back to this

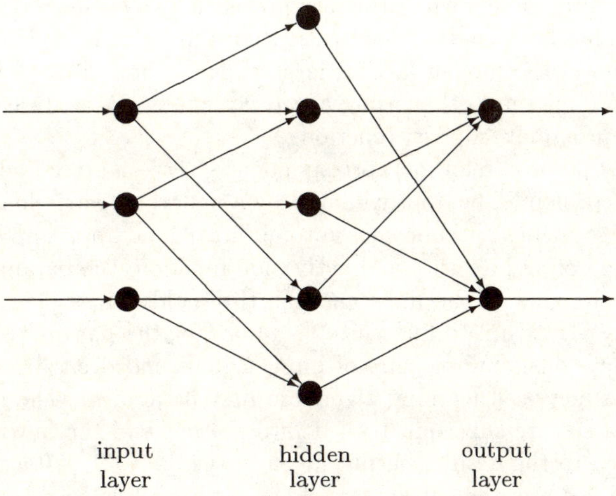

Fig. 2.4. A feedforward network with one hidden layer (a 3-5-2 network)

neuron again. Due to this structure the architecture of such a net is typically characterized as a layered structure. There is the input layer consisting of all nodes absorbing the signals from the environment, there is the output layer sending signals to the environment, and there may be some layers in-between consisting of neurons which process the signals from nodes in the previous layer in order to produce a signal which is sent to neurons in the next layer. Thus, the signal stream goes from the input to the output layers without any loops in-between. The layers of computation nodes between input and output layers are called hidden layers. The architecture of a NN is often characterized by the number of hidden layers in the net and the number of nodes in every layer. For example a single-layer feedforward network is a network consisting only of an input and an output layer but no hidden layers in-between. Networks with hidden layers are called multilayer feedforward networks. In figure 2.4 a network with one hidden layer with 3 input nodes, 5 hidden nodes and 2 output nodes (a 3-5-2 network for short) is shown. If every node in a layer sends its signal to every node in the next layer the network is called fully connected, otherwise partially connected. For example the network in figure 2.4 is partially connected.

Recurrent networks, on the other hand, must have at least one feedback loop. This means that a signal emitted by a neuron influences some later input of the same neuron. There may even be a self-feedback loop where the output of a neuron is fed back to its own input.

From a mathematical point of view a neural network with given architecture, weights, thresholds and activation function just represents a non-linear transformation of the input signal. It was shown that any function of bounded variation may be approximated with arbitrary precision by a feedforward network with one hidden layer and sigmoid activation functions provided that the number of units in the hidden layer is large enough (Hornik et al. [90]). Of course, this does not imply that using a two layer network is always the efficient way to approximate a given function.

The knowledge that any (relevant) correspondence between input and output signals can, in principle, be approximated by a neural network does not tell us how the parameters should be chosen to obtain a good approximation. Given a certain type of architecture and activation function, the parameters of the net are the weights of the links and the threshold values. Thus, we need some learning algorithm to find suitable values for the parameters. In case that a data set consisting of pairs of input signals and desired output exists, a so called supervised learning algorithm may be applied. The input vectors of the data set are subsequently fed into the net and the output of the NN is compared to the desired output given in the data set. Afterwards the error is determined and the parameters of the net are adjusted according to the input signal and the error signal. There are several learning algorithms for this adaptation, the best known maybe the back-propagation algorithm, but I cannot present them in detail here. A very good coverage of these algorithms can be found in Haykin [78]. The basic idea of all these algorithms is to adapt the weights and thresholds in a way to minimize the deviation from the desired output. After performing with the training data the network should be able to capture the basic structure of the environment.

In some applications no training data is available but there is a scalar performance index called the reinforcement signal depending on the output of the network. In such cases so called reinforcement learning is applied to adapt the parameters of the network in a way that an optimal reinforcement signal is obtained for any input signal. As the tendency to produce actions which lead to a high reinforcement signal is strengthened, the prospect to produce actions with only small reinforcement is weakened. Again, there are a number of different algorithms for reinforcement learning, like Q-learning or TD(λ) learning (see Haykin [78], Keerthi and Ravindran [95]).

If no training data is available unsupervised learning has to be used. In such a learning algorithm the neural network just looks at the data it is presented with, finds out about some of the properties of the data set and reflects these properties in the output. To perform unsupervised learning a competitive learning rule may be used. In using such a rule every layer is seen as a competitive layer consisting of units which compete to become active. In the simplest case there is only an input layer and a competitive output layer. A signal is fed to input units and their signals are transferred to the output units. Now the output unit with the highest level of internal activity becomes active and the vector of weights w_k of the winning unit k is shifted towards the input pattern. In this way the neurons become feature detectors.

Only this rudimentary sketch of the way neural networks work is given here, but there is a large number of books providing a foundation of neural network. For example to Haykin [78] or Hertz et al. [81] or White [183] provide a more detailed coverage of the topic than is possible here.

If an artificially intelligent agent is represented by a neural network, similar observations as in the case of a classifier system apply (there is also some formal analogy between CS and NN, see Smith and Brown Cribbs [161]). The form of the reaction function of the agents is not clear a priori and the crucial point again is the interpretation of the input and output of the network. If NNs are used to simulate the behavior of economic systems, in general no experimental data exists which could be fed to an artificially intelligent agent. This implies that the best suited learning algorithm is reinforcement learning with the payoff as reinforcement signal. Several examples of this kind of application of NNs are presented in section 2.4. Of course, NNs have been employed in many economic applications besides the simulation of an economic system (see e.g. Rehkugler and Zimmermann [140]).

2.3.4 Cellular Automata

Cellular automata (CA) were introduced by Von Neumann [174] in his pioneering work on self-organizing behavior. It consists of a regular uniform lattice with a discrete variable at each site, also called cell. The state of a cellular automaton is completely specified by the values of the variables at each site. A cellular automaton evolves in discrete time steps, with the value of the variable at one site being affected by the values of variables at sites in its "neighborhood" on the previous time step. Of course in principle a longer memory than one period could also be used, but such models are very rare. The neighborhood of a site is typically taken to be the site itself and all immediately adjacent sites. The variables at each site are updated simultaneously, based on the values of the variables in their neighborhood at the preceding time step, and according to a definite set of "local rules".

Initially, CA were regarded as mathematical idealizations of biological or physical systems in which space and time were discrete and physical quantities took on a finite set of discrete values. However, also several economic applications of CAs have been worked out.

Cellular automata may be defined in any dimension but let us first concentrate on one dimensional CA. In this case, the sites are ordered on a straight line. Thus, every cell has two immediate neighbors and in the most elementary non-trivial cellular automata the value of a cell in the next period depends on the current values of the cell itself and its immediate neighbors. The rule, determining which cells influence the values of a cell, is often called the "neighborhood structure" of the automaton. We call an one-dimensional automaton where the variables at the sites may only attain two different values and where the neighborhood of a cell consists only of the cell itself and the two immediate neighbors an elementary cellular automaton. A CA

state	1	1	1	1	0	0	0	0
of the	1	1	0	0	1	1	0	0
neighborhood	1	0	1	0	1	0	1	0
resulting value	0	1	0	1	1	0	1	0

Fig. 2.5. Example of a local rule of a cellular automaton. According to this rule the value of the central cell is the exclusive or of the previous values of the direct neighbors. The short name of this rule is rule 90 or modulo 2 rule

is called legal if the rule determining the value of a cell in the next period satisfies the following two properties. First, in a legal automaton where all sites initially have value 0 they must keep value zero. This implies for an elementary CA that a neighborhood of 000 has to yield value 0 in the next period. Second, a rule in a legal automaton has to satisfy "reflection symmetry". This means that two neighborhoods of a cell which are symmetric with respect to this cell must yield the same value of the cell. For elementary automata this implies for example that 100 and 001 lead to the same value of the middle cell. Further, a rule is called totalistic if the subsequent value of the central cell depends only on the sum of the values of the variables in the neighborhood. A well known example for a totalistic CA is the "Game of Life" (see Gardner [67] - [70], Berlekamp et al. [21]). Considering only elementary automata there are $2^3 = 8$ different possible states of the neighborhood and thus 2^8 different possible rules. A convenient way to write down a rule of an elementary CA is illustrated in figure 2.5. The rule is determined by a binary string of length 8, where the value at the i-th position gives the value of the cell in the next period if the current state of the neighborhood is given by the binary representation of i. For example, an 1 in the sixth position of the string means that a neighborhood of 110 yields of value of 1 for the central cell in the next period. With a neighborhood of 110 we mean that the left adjacent cell has value 1, the right adjacent cell has value 0 and the variable at the site itself is 1. Often a rule is identified by the integer equivalent to its 8 bit binary string. For example the rule of figure 2.5, which is quite popular in the literature,is called rule 90. The next value of a cell according to this rule is just the exclusive or of the previous values of its direct neighbors. As this can be expressed as the sum modulo 2 of the values of the direct neighbors, this rule is also often referred to as the "modulo 2 rule". Every legal rule must have the representation $\alpha_1\alpha_2\alpha_3\alpha_4\alpha_2\alpha_5\alpha_4 0$. This implies that only $32 = 2^5$ out of all possible 256 rules are legal. Besides, in only 4 of this 32 legal rules the value of the central cell does not depend on the last value of the central cell. These rules are the rules 0,90,160 and 250. On the other hand, any totalistic legal elementary automaton has a rule of

0	0	1	1	0
1	1	**0**	1	0
0	**1**	**0**	**1**	0
1	1	**1**	0	0
0	1	0	1	0

0	0	1	1	0
1	**1**	**0**	**1**	0
0	**1**	**0**	**1**	0
1	**1**	**1**	**0**	0
0	1	0	1	0

Fig. 2.6. The Von Neumann and Moore neighborhood of a site. The cells with bold faced values are in the neighborhood of the central cell. The left picture shows the Von Neumann neighborhood, the right one the Moore neighborhood

the form $\alpha_1\alpha_2\alpha_2\alpha_3\alpha_2\alpha_3\alpha_3 0$. Obviously $2^3 = 8$ such rules exist. Many further characterizations of the rules could be discussed but I will not go into details here and refer to Wolfram [185] for a good reference.

Of course, all these considerations may be extended to cellular automata of a higher dimension. For two dimensional automata, the two most popular neighborhood structures are the "Von Neumann neighborhood" and the "Moore neighborhood". Whereas the Von Neumann neighborhood contains besides the central cell itself only the four cells directly above below and besides the central cell, the Moore neighborhood contains the central cell and all eight cells adjacent to it. An example of both structures is given in figure 2.6.

There may be a finite or an infinite number of cells. If the number of cells is finite, the lattice is often closed to a ring in one dimensional automata or to a torus in two dimensional ones. In this case no cells are on the boundary. Otherwise, some special rules for the cells on the boundary have to be formulated.

A cellular automaton is an algorithm which, although based on very simple local rules, may generate quite complex overall behavior. It has also been observed in a lot of applications that CA may have a strong tendency towards self organization. This tendency is often explained by the irreversibility of the evolution of CA. Several constellations may lead to the same subsequent state of the automaton, which implies that the number of possible constellations decreases during a run. To deal with the global properties of a cellular automaton in a systematic way, Wolfram has proposed the following division into four classes which has become standard by now (see also Wuensche and Lesser [188] for a good survey of global behavior of one-dimensional automata):

1. Type I cellular automata have a unique steady state configuration, which is reached regardless of the initial state of the automaton. Thus, the effect of changing the variable value at one site completely disappears over time.

2. In type II automata evolution leads to a set of stable or periodic structures that are separated and simple. In this type of automata the effect of a local perturbation of the initial condition remains localized around the cell initially perturbed.
3. The class of type III automata is the most interesting one. These automata show chaotic aperiodic behavior (of course only in infinite automata; in finite automata the behavior has to become periodic eventually) and their evolution leads to "strange" attractors. Many of these automata exhibit self organizing behavior. Although a local perturbation of the initial configuration has long lasting global effects (sensitive dependence on initial conditions), the statistical properties of the long run behavior are the same for almost all initial conditions. The long run behavior of a type III cellular automaton is therefore governed by these statistical properties. The change of one single cell propagates to neighboring states at a fixed speed.
4. Evolution of type IV automata may lead to complex structures. Contrary to type III automata the long run behavior of such an automaton does not exhibit self organizing behavior, and, in general, the statistical properties of the long run behavior depend on the initial condition. Type IV automata are more rare than the other types and do not occur for elementary cellular automata.

As already mentioned, the type III automata are the class of CA studied most extensively in the literature. A lot of work has been devoted to analyze the complex patterns generated by type III rules. The key mathematical concepts for this kind of study are the Lyapunov exponents or the entropy of the system. See Wolfram [185] for an introduction to this topic. As already mentioned also rule 90 described in figure 2.5 is a type III rule. To illustrate the self organizing behavior of such an automaton, figure 2.7 shows the evolution of the corresponding CA starting from an initial constellation where only one cell has value 1. It can be seen that already after the 8 periods depicted self similarity occurs. The cells with value 1 build a large outer triangle which contains three small triangles of the same shape. With increasing time more and more of such triangles occur in different sizes. Finally, I would like to mention that automata belonging to class IV may have the property of computational universality, which means that these automata can evaluate any computable function (Wolfram [185], p.62f). For example the Game of Life is a type IV automaton which has been shown to be a universal computer.

For economic modeling CA are of special interest, because they are a very natural model of local interaction between economic agents. In many situations it is a very plausible assumption that agents interact only with individuals in their direct neighborhood. Think for example of an individual which imitates some other individual. Such an individual will in general adopt the strategy of another agent, which is close to him, so that he can observe the others' actions. It is also very plausible that a firm in a market

```
              1
            1 0 1
          1 0 0 0 1
        1 0 1 0 1 0 1
      1 0 0 0 0 0 0 0 1
    1 0 1 0 0 0 0 0 1 0 1
  1 0 0 0 1 0 0 0 1 0 0 0 1
1 0 1 0 1 0 1 0 1 0 1 0 1 0 1
    ⋮           ⋮           ⋮
```

Fig. 2.7. The evolution of a rule 90 automaton starting with an initial condition with only one cell with value 1. As well the global effects of the different value of this single cell as the self organizing behavior can be seen quite clearly

has only information about the output of some firms it is in contact with but not about all firms in the market. As such spatial structures are of great relevance in economic systems, CA have recently been used in quite a number of economic models. Normally, in these models every site represents one individual and the possible states of the site represent the different actions at his disposal. The rule of the automaton models the adaptation of the agent in reaction to the actions of the surrounding individuals. Note, however, that contrary to models where classifier systems or neural networks are used for modeling artificially intelligent agents, the reaction function of the agents is given a priori. So no adaptation of the decision making process per se occurs but the evolving behavior of locally interacting individuals with given behavioral patterns is studied. In the examples presented in the next section it will become obvious that also in economic systems type III automata have attracted the most interest and that complex patterns may occur in spatial economic systems.

Whereas the explicit modeling of a spatial population structure is the main advantage of this approach, it could be argued that a major drawback lies in the fact that such kinds of automata do not converge to a stationary state, but keep on changing till they may sometimes end up in a circle. A consistent analysis of such behavior seems to be quite difficult, for the attributes of the population may change every period. However, for automata of type III or lower the proportion of the different strategies will remain more or less constant after some time, which implies that also the evolution of a cellular automaton may be interpreted as the learning of a certain population state.

2.4 Some Applications of CI Methods in Economic Systems

Having sketched the basic ideas of some of the most popular CI techniques in the last section I will now shortly describe some examples where these techniques have been applied to simulate the behavior of economic agents. In the last couple of years the number of articles published in this area has exploded and it is impossible to provide even a rough overview here. However, I think that it is helpful to look at some selected applications in order to see the success but also the shortcomings of the approach taken here. Further contributions to this field can for example be found in Biethan and Nissen [26], Conte et al. [39], Moss and Rae [124] or special issues of the *Journal of Economic Dynamics and Control* (Vol. 18, No. 1, 1994; Vol. 21, Nos. 8-9, 1997), *Computational Economics* (Vol. 8, No. 3, 1995; Vol. 9, No. 1, 1996) or the *Journal of Economic Behavior and Organization* (Vol. 29, No. 1, 1996). See also the overview in Holland and Miller [89] and the book by Epstein and Axtell [59]. Alander [2] provides a rather extensive and regularly updated bibliography of Genetic Algorithm applications in economics[2].

2.4.1 Bidding Strategies in Auctions

One field where artificially intelligent agents were used successfully to imitate a population of adaptive human agents is the building of trading strategies in auctions. Andreoni and Miller [6] have done a large number of simulations, where genetic algorithms were used to represent the evolving trading strategies of the bidders. They did simulations in the case of first price and second price, common value and private value auctions [3]. The parameters of the Nash equilibrium bidding function were coded as a binary string and the genetic algorithm worked on the population of these strings. The simulations showed that in all cases the genetic algorithm was able to come close to the Nash equilibrium. They observe that the artificially intelligent agents exhibit many of the same bidding patterns as human agents do. In particular, the trend to overbid that can be observed in experiments with human bidders, showed also in these simulations with genetic algorithms. The authors notice further that the learning of the Nash equilibrium seems to be easier participating in an environment of evolving strategies than confronted with a static environment of Nash strategies. This result seems to be rather surprising as non equilibrium strategies may be strengthened in the case of a population of evolving strategies, whereas in the environment of Nash equilibrium strategies the task is only to find a best reply to the given environment. They claim however that the evolving population leads to a better combining and sharing of information.

[2] Available via anonymous ftp at ftp.uwasa.fi/cs/report94-1/gaECObib.ps.Z.

[3] See Andreoni and Miller [6] for a detailed description of these different types of auctions.

Whereas Andreoni and Miller used the computational intelligence technique merely to find the equilibrium values of some parameters in the equilibrium bidding strategy, Rust, Palmer and Miller [145] go one step further and represent the whole bidding strategy by a neural network. The three researchers held a computer tournament for double oral auctions at the Santa Fe Institute in 1990 and 1991 (see Rust, Palmer and Miller [146]). 30 bidding strategies implemented as a computer program took place in this tournament. The relatively simple winning strategy was submitted by Kaplan and can be characterized as "*wait in the background to let others do the negotiating, but when bid and ask get sufficiently close jump in and steal the deal*". After this tournament of "human" bidding strategies Rust, Palmer and Miller tried to find even better bidding strategies by using artificially intelligent agents. They represented each bidding strategy by a neural network. The neural networks were trained by competing against each other, but also by competing against other strategies. Overall there were 52 buyers and 52 sellers, where some of the buyers and some of the sellers were represented by neural networks. The simulations were done over a long period and showed that the neural network traders were only able to outperform Kaplan's strategy in the beginning of the simulation. After some time the success of the neural network strategy decreased strongly and was clearly beaten by Kaplan's strategy in the end. Although there is, of course, some probability that another network architecture than the one used by Rust, Palmer and Miller could lead to better results, we have to accept that in this setup the computational intelligence techniques were not able to gain the results earned by human decision makers. We will analyze adaptive learning in an auction market more thoroughly in chapter 6.

2.4.2 The Iterated Prisoner's Dilemma

One of the most widely used models in order to illustrate the evolution of cooperation in socio economic systems is the simple and highly stylized evolutionary game called prisoner's dilemma. The payoff matrix of the game is given by

$$A = \begin{pmatrix} (R,R) & (S,T) \\ (T,S) & (P,P) \end{pmatrix}, \quad T > R > P > s. \tag{2.1}$$

In the literature the first of the two pure strategies is called "cooperation"(C), whereas the second is called "defection"(D). The story behind the payoff matrix is the following: Two people have together committed a major crime and are caught afterwards by the police. However, the police is only able to prove another minor crime of the duo. The sentence for this minor crime is only a short stay in prison. If one criminal witnesses against the other, he will go free and the other will be sentenced to a long prison term. If however both like to witness against each other, both have already admitted the crime and therefore both will be sentenced to a jail term which is longer than the one for the minor crime. The story is a little far fetched, but there are many situations in economics that can be modeled in a similar way (think for example of a duopoly). It is obvious from the payoff matrix that D strongly dominates C, which implies that (D,D) is the only Nash equilibrium of the game. On the other hand, (D,D) yields for both players a lower payoff than the state of cooperation (C,C). It is however easy to verify that if the prisoner's dilemma is played repeatedly between the same opponents, where the time horizon is possibly infinite, there are Nash-equilibria consisting of strategies that imply mutual cooperation.

In 1979 Axelrod organized a computer tournament where different game theorists submitted strategies to play the iterated prisoner's dilemma (IPD) (see Axelrod [15]). 15 strategies were involved in the contest and every strategy played against all other over a sequence of 200 moves. The winning strategy was submitted by the economist Anatol Rapoport and became famous under the name "Tit-for-Tat". The strategy is very simple and can be characterized as follows: Cooperate on the first move and do whatever the opponent did in the last move afterwards. The main advantage of this strategy is that it will always cooperate when it is matched with a nice strategy, but can on the other hand not be exploited by nasty behaving defecting strategies. In 1980 Axelrod organized a second tournament for the IPD. This time 62 strategies were submitted, and although all these strategies were developed with the knowledge that Tit-for-Tat has won the first tournament, also this second tournament was won by Tit-for-Tat. Tit-for-Tat was inferior to almost all strategies in the direct contest, but the new strategies earned very little in the contests against each other, so that Tit-for-Tat could again gain the highest average payoff in the tournament. Axelrod observed that there were eight special strategies in the second tournament, that could be well used to calculate the payoff of a given strategy in the whole contest. The average payoff against these eight strategies was in many cases very similar to the average payoff against the whole population.

Axelrod used these eight strategies as an environment for a genetic algorithm in order to learn strategies in the IPD (Axelrod [16]). He coded strategies which use the last three moves to determine the current move. This means that there are 4^3 possible histories considered by the players and a strategy has to determine the current move for any of these histories. Axelrod used quite a simple coding by numbering the possible histories from 1 to 64 and stating that the value of the bit in position i determines whether to cooperate or defect whenever history i occurred. Additional six bits were used to encode a fictitious history to be used in the beginning of the iterations. Therefore, Axelrod used binary strings with length 70 to represent a strategy in the IPD. In figure 2.8 an example of a string encoding a strategy in the IPD is shown. Due to space constraints not all 70 bits are written down but the structure can be seen. The value "1" should be identified with "cooperate", "0" with "defect". A history of "CDCCCD" should be read as

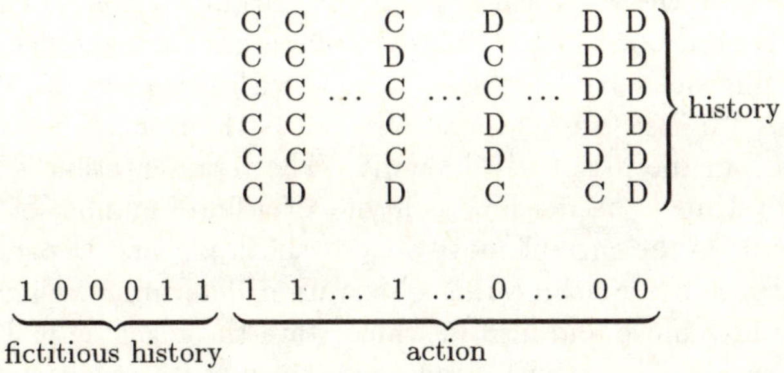

Fig. 2.8. An example of the Axelrod like encoding of a strategy in the IPD. The action part of the string contains 64 bits

follows: three periods ago the opponent cooperated and the player himself defected, two periods ago both cooperated and in the last period the opponent cooperated and the player himself defected. The strategy depicted here could for example be Tit-for-Tat but of course also many other strategies. Of course, if all 70 bits were given this would define a unique strategy. Note that in the first period the strategy is chosen according to the fictitious history of "CDDDCC".

In Axelrod's simulations the population size was 20, and the fitness of the strings was calculated by opposing them with the eight representative strategies of the second tournament. The simulations showed that the genetic algorithm was able to create populations of strategies whose average payoff was as high as the average payoff of Tit-for-Tat. Many of the generated strategies resembled the behavior of Tit-for-Tat. In other experiments Axelrod did not use a static environment of eight representative strategies to calculate the fitness of a string, but opposed each strategy with the other strategies in the evolving population. These simulations led in some cases to the generation of even more successful strategies than Tit-for-Tat. These strategies were in most cases no nice strategies and started with defection in the first move. As in the case of the bidding strategies in auctions, this implies that an evolving population will probably yield better results when the fitness is determined by contests within the population, rather than by opposing the strategies to a given external environment.

Mühlenbein [126] extended Axelrod's work by addressing the question whether a spatial population structure improves the performance of the genetic algorithm. Relying on the *continent cycle theory* of Darwin [42] he performed simulations where the population is periodically split into islands and the small island populations evolve separately for some time until the population is reunited. The simulations show that a population subject to such a cycle will arrive at cooperation more rapidly than any other used population structure, and will earn the highest average payoff.

In addition to the work using genetic algorithms to generate good strategies in the IPD also a number of other techniques from the field of computational intelligence have been used to cope with this problem. For example Fogel [60] used finite state machines also called Moore machines to represent the strategies in the prisoner's dilemma . The representation is straightforward. At any time t the machine is in one of a finite number of states. The machine observes the current move done by both players. Dependent on the current state of the machine this observation determines which action to take in the next move and also in which state the machine will be at time $t+1$. Fogel used machines with a maximum of eight states and used standard techniques from the theory of Evolutionary Programming (EP) to evolve a population of such machines. Note that there are strategies which can be encoded by such a machine but not with Axelrod's encoding but on the other hand also some of the strategies which may be represented in Axelrod's setup can not be described by an eight-state machine. Fogel did simulations with 150 iterations and especially analyzed the impact of the values in the payoff matrix of the game on the evolution of cooperative strategies in the population of machines. He showed that cooperative strategies emerge in this setup only if the payoff of mutual cooperation is higher than the payoff expected by a player that alternates between cooperation and defection (i.e. $2R > T+S$). In cases where this relation does not hold, machines evolve that will often end up alternating between cooperation and defection. Fogel observes, therefore, that cooperation will reliably evolve in real world circumstances, if the reward of defection against cooperation is small relative to the net benefit of mutual cooperation. The strategies generated by the described procedure were in some cases quite complex and rather difficult to interpret. In cases where the payoff of mutual cooperation is large enough the machines evolve towards a cooperating population, but it may nevertheless happen that machines generated in two different such simulations will not cooperate when battling against each other. The explanation given by Fogel for this rather interesting fact is that the machines build patterns which indicate to other machines whether this machine will respond to cooperation with cooperation or defection. Obviously these patterns are incompatible if the machines developed in different populations, and this may prevent the machines from mutual cooperation.

Based on the paper of Fogel, Hanne [78] did simulations in similar settings but this time with a genetic algorithm modeling the learning process. His results are quite similar to these obtained by Fogel, particularly they also show that the emergence of cooperation depends on the reward of defection. The fact that this observation has been made with different adaptive processes, shows that the result should be generic and not depend on a special learning process.

Ho [84] also used GAs to update a population of Moore machines representing strategies in the IPD. He shows that penalizing complex strategies based on the size of the machine can destroy cooperative behavior. On the other hand, penalty costs which increase with the frequencey of state switches does not hurt the development of cooperative behavior.

Another approach to model the evolution of strategies in the iterated prisoner's dilemma was chosen by May and Nowak [118, 119] and Kirchkamp [96]. They assume that the individuals are ordered on a line or a rectangular cluster and that each individual interacts only with other individuals that are nearby[4]. May and Nowak choose a setup where every individual has only a simple one step strategy that says whether to cooperate or defect in the next period regardless of the last move of the opponent. They assume further that a player always switches to the strategy of the player in his neighborhood who does best. This setup defines a two dimensional cellular automaton, where each individual is represented by a cell and the state of the cell determines which strategy to play in the next move. Each individual plays the prisoner's dilemma against all of his eight neighbors which yields his payoff. Afterwards the individual adopts the strategy of the neighbor with the highest payoff. This implies that the subsequent value of the central cell depends on the previous values of the eight neighbors and their neighbors which are altogether 25 cells. Thus these 25 cells are the neighborhood of the central cell. The authors demonstrate that this CA may exhibit a large variety of behavior. Besides static patterns complex structures like spatial chaos, dynamic fractals and kaleidoscopes are observed.

Kirchkamp [96] goes one step further and uses two-state Moore machines to represent the strategies of the players. This means that every cell may be in 26 different states, one for each machine with less than three states. Whereas the simulations, in the case of simple one-step strategies, show that a population of defectors will evolve as long as the payoff of mutual cooperation is significantly lower than the payoff of defection against cooperation, in the case of discriminatory behavior (two state machines), the average payoff in the evolved population is always near the Pareto frontier, given by the payoff of mutual cooperation. However, there evolves no uniform population consisting for example only of Tit-for-Tat machines, but the population keeps heterogeneous and strategies like Blinker, which always alternate between cooperation and defection, stay quite strong in many cases.

2.4.3 Goods Markets

The modeling of markets are an especially promising field for the application of of CI techniques in economics. Several papers have studied the interaction of artificially intelligent agents in a computer simulated market place. The crucial question is again whether the individuals or firms represented are able to develop efficient behavioral patterns in a market situation.

[4] In an interesting recent contribution Berninghaus and Erhart [23] study the evolution of local interaction *structures* in games using Genetic Algorithms.

Arifovic used genetic algorithms to analyze the behavior of adaptive agents in a number of simple economic standard models [8, 9, 10, 11, 12]. In Arifovic [8] she used a simple cobweb model as the economic environment. The model assumes that there are n competing firms that are price takers and produce the same good. The firms have quadratic costs and the demand function is linear decreasing in the average produced quantity. Each firm is faced with the problem to choose the optimal quantity to produce without knowing the market price of the good in the current period.

In the GA every firm is represented by a binary string that encodes the quantity the firm produces in the current period. In addition to the operators used in a simple genetic algorithm Arifovic used also an additional operator called election, which will be explained further in the next chapter. Using this additional operator allowed the GA to learn the rational expectations equilibrium values for the cobweb model. She noticed also that the GA was able to learn the equilibrium in cases where analytical approaches like least squares learning diverge.

Franke [64] extended this analysis by assuming that the demand curve is perturbed by random shocks following an AR(1) process. The strategies of the market members are not necessarilly the production quantities but the agents might also use adaptive or regressive expectation schemes in order to determine their output. It turns out that robust convergence towards the equilibirum can be observed also in this framework and that in general a large part of the population uses a simple linear adaptation scheme in order to build their price expectations. The framework of a cobweb model will also be used in chapter 6 of this monograph.

A spatial market model was presented by Albin and Foley [3]. They carry out simulations of exchange among agents which are located on a circle. There are two goods in the market. All individuals have the same utility function, and the structure of this function implies that each agent tries to consume the same amount of both goods. If his current endowment does not consist of the same amount of both goods he is willing to exchange some amount of one good against some other amount of the other good. Thus he has a marginal rate of substitution of good 2 for good 1 which is not equal to 1. An individual willing to exchange good 1 for good 2, or the other way round, may make his willingness public by advertising it. Advertising reaches only some neighborhood of the agent and causes costs proportional to the size of the neighborhood reached. On the other hand, it is necessary to find a partner for exchanging goods. When a buyer and a seller of the same good have both advertised they exchange some amount of goods if their marginal rates of substitution are compatible. The expected gain from advertising depends on the expected probability that some other individual will advertise a signal leading to a deal in the next period and this probability is estimated statistically from data gathered in the own neighborhood. Thus, formally the model may be regarded as a one dimensional cellular automaton with a continuum of possible values for the variables at each site. The simulations show that with random initial endowments, after some periods of heavy advertising and trade, the market reaches a state where the marginal rates of substitution of adjacent agents differ only slightly. Although, there are still agents whose marginal rate of substitution lies considerably above and others whose marginal rate of substitution lies considerably below zero, no advertising is done because these agents lie too far apart to reach each other by advertising. The authors determine by simulations that the best resource utilization of the whole system appears if the neighborhood of a cell is defined by the r close neighbors of the cell on either side, where r is about five percent of the whole number of cells in the system.

2.4.4 Financial Markets

The SFI artificial stock market developed at the Santa Fe Institute is a very good example how artificially intelligent agents may be built to mimic certain important qualitative propertites of real life traders. Arthur et al. [13] describe the strucutre of the market and also compare the outcome of the artificial market with real live stock data. The market model is simple with a single risky assets whose dividend follows an AR(1) process and a risk free bond. A trader in the market is represented by a classifier system which determines the traders expectation about price plus dividend of the risky asset in the folloing period. It is always assumed that the expectation function is linear and the action part of each rule gives the slope and position of the function. The condition part, on the other hand, is binary and is matched with a description of the market, where each bit in the string describes a certain aspect of the state of the market. For example, a certain bit is one whenever the current price is larger than the 5-period moving average of past prices and zero otherwise, another one is one whenever the current price is above the fundamental value. These bits can be interpreted as 'technical' or 'fundamental' descriptions of the market and such an interpretation allows interesting insights whether fundamental or technical traders prevail in the artificial stock market. The traders calculate their optimal demands based on these expectations and afterwards the market clearing price is determined. The strengths of the rules in the classifier system are updated using the forecasting errors made when active. The main results reported by Arthur et al. [13] may be summarized as follows. If learning is slow (the GA is invoked only every 1000 period and the updating of a predictors accuracy is slow) always the rational expectations equilibrium is reached. With faster learning, however, complex patterns with price bubbles and crashes emerge. In this case – contrary to the slow learning case – it turns out that the technical trading bits are of high importance for the prediction of future prices plus dividends. Several statistical properties of the time series in the complex scenario were compared to those of IBM stock data and found to be qualitatively similar.

An artificial stock market populated by three types of neural network-based agents was analyzed by Beltratti and Margarita [19]. One type of share is traded in the market and based on previous values of the price and change of price of the shares the agents build expectations about the price of one share in the next period. The three types of agents are denoted as smart, dumb and naive. The smart and the dumb agents are both represented by a feedforward neural network with one hidden layer and back-propagation learning. However, due to a special constellation of weights the dumb agents always predict an extreme value of the price (0 or 1). On the other hand, the network representing a naive agent consists only of one input and one output unit without any hidden unit inbetween. The naive agent bases his expectations only on the previous market price. Contrary to the other two types the naive expectations cause no costs. Agents meet randomly and the individual with the higher price expectation buys one unit from the one with the lower price expectations, where the actually paid price is the mean value of both price expectations. After every period the weights in the networks are updated with the help of observed data. Every T periods the agents may choose a new strategy. The probability that a certain strategy is adopted is proportional to the wealth of the agent who previously used this strategy. Afterwards the strategy is mutated randomly. Simulations show that the dumb agents always disappear quickly. Initially there is a high percentage of smart agents, but as the market stabilizes also naive expectations are good and the naive agents who do not have to face costs for building their expectations may take over the population if the expectation costs of the smart agents are too high. On the other hand, the introduction of exogenous shocks prevents such a development. An especially interesting aspect of this model is that the population is heterogeneous in the sense that different agents use different types of learning rules.

Lettau [109] studies the portfolio decisions of boundedly rational traders using Genetic Algorithms. His model is simpler than the ones mentioned above in the sense that he assumes that the price of the risky asset is not determined in the market but given exogeneously. He performs a thourough comparision of the data obtained in numerical experiments in this framework with empirical data. I will get back to the results of this comparision in the next chapter.

2.4.5 Evolving Market Structures

The question how decentralized trade orgnaizes if no central market place with market clearning is assumed is hard to analyze using analytical tools. However, the use of artificially intelligents agents enables us to study under which circumstances self organization of markets inhabited by boundedly rational individuals can be observed. Vriend [178] considers a decentralized market where some of the agents are producers and others consumers. Consumers try to get one unit of the single good in the market, however they cannot distinguish between producing and not producing agents in the market before they shop around. Since they can only visit one other agent a day visiting an agent who did not produce that period or has already sold all units causes dissatisfaction for that date. Thus, it is important for the consumers to have a good strategy on how to decide which agent to visit in some given period. On the other hand, a producer has to decide how much to produce and whether to send signals to other agents indicating that he will sell the good in the current period. Both activities are assumed to be costly. The agents are represented by classifier systems. The results indicate that the market organizes very quickly in a sense that after a very short learning phase practically all consumers manage to buy a unit of the good every period but, on the other hand, the signalling costs in the market are very large compared to actual production costs. This inefficiency could be avoided if the behavior of all consumers could be exogeneously influenced such that they always revisit an agent if they were able to buy the good from him the previous period. This demonstrates again that the non-cooperative nature of agent based models of this kind may induce certain effects in the population which cannot be obtained in a representative framework[5].

The analysis of emerging market structures is also the topic of Kirman and Vriend [98]. Their model describes the transactions on the wholesale fish market in Marseille. The problem of the buyers in this markets is on one hand to decide which seller in the market to visit and on the other hand to decide which prices to accept. The sellers have to determine their amounts of supply and asking prices but also how to treat customers who have already been buying from this seller in previous periods. The agents are again represented by one classifier system each. The main findings of this model are that price dispersion exists in this artificial market and that 'loyalty' emerges. Buyers learn to become loyal to certain sellers and the sellers learn to offer higher payoffs to loyal buyers. Both observations could also be infered as stylized facts from the real fish market in Marseille.

An interesting market model with explicit spatial strucuture was analyzed by Keenan and O'Brien [93]. They consider a fixed number of firms equally spaced along a circle. Consumers are located symmetrically on the circle and purchase goods from their neighboring firms. There are only two possible actions for the firms: setting a high cartel price $\tilde{p}$ or a low competitive price $\hat{p}$. There are two kinds of consumers in the model. One type always buys from the firm with the lower of the two prices, whereas the second type buys from the firm nearest to him, regardless of the price. Each period every second firm reestimates its price decision. Thus adjacent firms set their prices every other period. The decision rule of the firms are all the same and very simple. They consider only the set prices of their two direct neighbors and

[5] Closely related to this approach is also the work on endogeneous partner selection in games and the evolution of trade nteworks (see Ashlock et al. [14], Tesfatsion [166]).

set a high price whenever both adjacent firms sell at the same price and at the low price if the adjacent firms sell at different prices. The authors argue that this kind of behavior may be implied by diseconomies of scale and the given structure of consumers. If we consider only the odd firms at the odd periods this model describes exactly an elementary cellular automaton with the modulo 2 rule (rule 90). The same holds true if we consider the values of the even cells at even periods. As pointed out in the last section, the rule 90 determines a type III automaton and leads to complex patterns and a high degree of self organizations. This effect can also be observed in this model. Starting with initially uncorrelated prices the spatial market develops towards a situation where cartels of adjacent high pricing firms emerge. Cartels of the same shape but different sizes appear regularly in different places of the space. Subsequently these cartels shrink and disappear as the firms at the edge of the cartel always change to the competitive price. Similar results with even more preserving cartel building could be observed in models with different reaction functions. As is to be expected, with these different reaction functions the resulting rules are different from the rule 90 but the automata are still of type III.

2.4.6 Further Simple Economic Models

Marimon, McGrattan and Sargent [113] inserted artificially intelligent agents in a simple model of money. The model which originates from Kiyotaki and Wright [99] assumes that the economy consists of a given number of different types of households. It is assumed that the types of households are ordered by a total ordering. Each type of household produces its own type of good, but is only interested in consuming the type of good produced by a household of a type one step higher than its own type (the households of the highest type like to consume the goods produced by households of the lowest type). Each household can store only one unit of a good at a time, causing storage costs. The households are matched randomly each period, and try to exchange goods in a way to finally get hold of their aspired consumption good. A stationary equilibrium in this model is given by a trading strategy for each type of household and a set of probabilities of meetings between households storing different types of goods. Marimon, McGrattan and Sargent represented each household by a classifier system and were able to watch their convergence to an equilibrium in a model with three different types of households. They were even able to find an equilibrium in a model with five different types of households. This equilibrium was not known before and is a good example to show that techniques from computational intelligence may not only be used to simulate the behavior of adaptive agents, but are also well suited to compute equilibria in very complex systems where the analytical computation of equilibria may be cumbersome.

Litterman and Knez [110] used the same model as given by Marimon, McGrattan and Sargent but represented each household by a population of binary strings governed by a genetic algorithm. Also this approach led to the learning of an equilibrium in the model with three different types of households.

In other simulations, Arifovic [9] used GAs also to learn stationary equilibria with perfect foresight in a simple overlapping generations model. This model will be presented in some detail in chapter 6 and her results which indicate that the GA learns the stationary equilibria quite easily in cases where no other equilibria exist will be extended by demonstrating that GAs are also able to learn cyclical or stochastic equilibria if the appropriate setup is used. Arifovic extended her analysis to the case of a two country OLG model in [10] and Bullard and Duffy [36] dealt with an n-period OLG model where the adaptive learning of the agents is again modelled by a Genetic Algorithm.

2.5 Potentiality and Problems of CI Techniques in Economics

As already mentioned, the examples of the application of artificially intelligent agents in economic systems given above are only a small part of the rapidly growing literature devoted to this area. They show that these techniques have been applied with success in a large number of different fields of economics. I will use these examples to discuss some advantages and disadvantages of the use of artificially intelligent agents in economic systems. Obviously one of the most desirable feature of the simulations with artificially intelligent agents is the explicit representation of every individual in the population. Contrary to the econometric learning rules it is basically possible to build a heterogeneous population of agents who do not only differ in their strategies, but also in their learning behavior. An example for such a population is given for example in Beltrati and Margarita [19] and we will see another one in chapter 5. This feature is very important, for it is by no means clear why different individuals which are assumed to act differently as a member of the economic system have to use the same rule in order to build expectations or update their strategies. Although, the opportunity to diversify the population with respect to the transition rules from one period to the next is seldom used in the simulations up to now, this could be a major advantage of this technique.

The explicit representation of every individual enables the researcher running the simulation also to gain a deeper insight into the building of order in the randomly initialized population. Modern computer technology enables us to watch the evolving population like a fast movie of the evolution, and therefore we may observe in great detail the learning behavior of the population. A good example are simulations where cellular automata are used and where due to the implemented topological structure of the population it is especially interesting to watch the evolving population. For example it can be easily seen where a certain strategy appears in the population, how this strategy spreads through the population, and which impact the occurrence of new strategies has. Insights of this kind are completely impossible to obtain from analytical learning models.

In considering the individual rationality assumptions used in these examples, we have to be aware of the fact that these assumptions vary from model to model and are no inherent notion of any of the techniques used. If we consider, for example, the learning of the rational expectations equilibrium in the cobweb model by genetic algorithms, we have various possibilities to interpret the binary strings representing an individual. The string may encode the quantity the firm will produce in the current period, or it may also encode the expectation of the firm about the price in the current period (see e.g. Franke [64]). Whereas in the first case no rationality assumptions at all are implemented, the second coding makes only sense if we assume that the firm is able to calculate its optimal output given a certain price in the current period. In other words, the second coding implies individual rationality of every agent and uses, therefore, the same rationality assumptions used by most of the econometric models. Notice however that the researcher will in many cases have the opportunity to abandon also the assumption of individual rationality, and in these cases the population will not only learn to make good predictions, but also to act optimal just by interacting with other individuals.

As another advantage of the approach used here, compared with analytical approaches, may be seen that all the techniques cited here do not need an analytically given payoff function. The payoff function may be a black box, as no characteristics of the function apart from the function values are used.

Besides all these nice notions of simulations with artificially intelligent agents we have to face, however, also a number of weak points of the approach. As the rest of this work is concerned with the analysis of the behavior of genetic algorithms in economic systems, I will restrict the discussion to this kind of algorithm, but most of the arguments will hold also for the other CI techniques. Considering GAs we have to realize that a literal interpretation of the learning rules is quite difficult, if we consider socio economic learning processes rather than real evolutionary processes. I will discuss possible interpretations of the different genetic operators in an economic environment in the next chapter and we will see that these interpretations are not completely satisfactory.

Another drawback which is a consequence of the last point is that there is no economic argument implying which values should be attached to the parameters, that govern the behavior of the GA. It is however well known that different sets of parameter values may lead to quite different results, which implies that the setting of the parameters is of great importance. Usually those parameters are chosen which have been successfully used in static optimization problems, but this seems to be a weak argument, for we are not interested in optimizing a function but in simulating the behavior of a population of adaptive agents. A possible solution to this problem may be to compare computer simulations with human experiments in very simple models, and calibrate the parameters according to the experimental data. In the next chapter I will shortly present some cases where this has been done and discuss the results. It is however questionable if the same parameter setting will also be suited to mimicking human behavior in other and perhaps more complex systems. Similar arguments apply to the problem of the coding of the actions of the agents. There is a lot of evidence in the GA literature showing that different coding procedures may lead to different learning behavior. Again no economic argument can be given to decide the problem, and this must be seen as another drawback[6].

Finally, a large drawback of the modeling of learning by using CI techniques is the fact, that almost no mathematical theory exists which explains the results of the simulations. It is therefore impossible to predict the behavior of a system like a GA without really doing the simulation. Of course there are a number of heuristic explanations for different results, but it is of great importance for the further development of this line of research that a mathematical theory is built which is at least in simplified models able to answer questions concerning the global or at least local stability of equilibria. As these results depend on the values of the parameters and the coding used, such a theory may also be used to determine which settings are plausible in certain environments and which are not. Further theoretical considerations may help the researcher to determine whether some limit point of the simulated system is an equilibrium of the underlying economic system or not. The obstacles that withstand such a project will become apparent in the next chapter. A genetic algorithm, just like all of the techniques cited above, gives rise to very complex systems, when explicitly written down as a mathematical model. The analysis of the system is especially difficult, since the evolving system is nonlinear and acts in a space of huge dimensionality. Nevertheless, we will see that some steps in the direction of a theoretical foundation of CI simulations in economics can indeed be done.

[6] Note that I am talking about the procedure that encodes a given number or strategy or automata in a binary string here. The interpretation of this number is another problem and can be decided on grounds of economic arguments as shown above.

3. Genetic Algorithms

3.1 What are Genetic Algorithms?

In the rest of this monograph I will deal exclusively with a special technique from the field of computational intelligence research, namely genetic algorithms. Genetic algorithms were developed by Holland [87] in 1975 as a tool to find solutions of optimization problems in poorly understood large spaces. They are based on the genetic processes of biological organisms, especially on the principle of natural selection that has become famous as "*survival of the fittest*" since the publishing of *The Origin of Species* by Charles Darwin [42]. Although, this slogan seems to be slightly tautological in the natural environment, where fitness is defined as the ability to survive, it makes good sense in the world of optimization problems, where the fitness of a string is given as the value of the function to be optimized at the argument encoded by the string. GAs proved to be quite successful in finding good solutions to such complex problems as the travelling salesman problem, the knapsack problem, large scheduling problems, graph partitioning problems, but also for engineering problems like the design of bridge structures or the optimal use of power plants. The most famous application is perhaps one of the first large applications done by Goldberg. By using classifier systems based on GAs he was able to generate a control system that managed the flow in the large gas pipeline from Texas to Chicago (Goldberg [72]). In 1989 Goldberg wrote a seminal book dealing with genetic algorithms, that is up to now probably the most widely spread book in this field of research (Goldberg [74]). Besides the principle of natural selection, GAs imitate not only the spreading of genetic material in a population but also the generation of new genetic material by mutations . Other features like the duality of genes or the existence of dominant and recessive genes are typically neglected in artificial systems. It is, however, up to now, by no means clear which processes in nature are essential for the process of evolution and which are not, so that the choice of the features that are transferred to GAs is due mainly to implementation problems. It is quite important to realize that an artificial GA is only a very simple model of the natural evolution, and that natural evolution is basically not an optimization technique, but it just evolves. I think, therefore, that the occurrence of some process in natural evolution is per se no argument that an operator modelling this process will improve the learning behavior of a

```
BEGIN
        create initial population P_0
        WHILE NOT stopping condition DO
                apply genetic operators to generate P_t from P_{t-1}
                check stopping condition
        END
END
```

Fig. 3.1. The structure of a Genetic Algorithm

GA. The connection between the natural and artificial evolution is not close enough, so as to guarantee that any argument valid in one world will also be valid in the other.

In order to facilitate the reception of the results derived in the following chapters, I give a short introduction to the structure of genetic algorithms and review the most important approaches to an analytical treatment of the behavior of GAs. Nevertheless, the interested reader should consult introductory literatures like Goldberg [74], Michalewicz [120] or Mitchell [122] for a more exhaustive coverage of the topic.

3.2 The Structure of Genetic Algorithms

Typically, a genetic algorithm works on a population of binary strings. These strings correspond to the chromosomes in natural systems. In the standard setup all strings have the same length which is one of the parameters that have to be set by the researcher implementing the GA. Let l denote the length of a string. Another parameter to be determined is the size of the population. This parameter is called n. The initial population – denoted by P_0 – is often generated randomly, but may also be set to a certain state, so as to watch the evolution of the system for a given initial condition. There has to be an externally given fitness function denoted by $f(\cdot)$, which is to be optimized. It is assumed that the fitness function is positive. In cases where the objective function of the problem happens to be negative, or where the objective is to minimize a given function, the fitness function will not coincide with the original objective function, but will in general be a transformation of this function. The transition from the population at time t denoted by P_t to P_{t+1} is executed by applying several genetic operators (a simple diagram showing the structure of a GA is given in figure 3.1). These operators may be quite specific in single applications, but there are a number of standard operators which constitute the so called simple genetic algorithm. I will shortly describe these standard operators in section 3.3. After the creation of P_{t+1}, the algorithm checks whether or not the stopping condition is met. If it is not, the genetic operators are applied to P_{t+1} to create the next population. The most widely used stopping conditions are either that a given number of iterations have been carried out already, or that the population has become uniform, which means that all strings in the population are identical. In this case we say that the algorithm has converged, although there is no convergence in the mathematical sense, for the mutations disrupt the uniform state over and over again.

3.3 Genetic Operators

3.3.1 Selection

The selection operator is intended to implement the idea of the "*survival of the fittest*". In optimization problems such an operator is the force that is responsible for the fact that a GA will on average climb up the function graph. Basically, the selection operator determines which of the strings in the current population is allowed to inherit their genetic material to the next generation. If we use the GA language we say that it builds up the mating pool by selecting n strings from the current population. The standard selection operator, called proportional selection, does this by carrying out n random draws with replacement out of P_t. The probability that a given member j of the current population is chosen at one certain draw is proportional to the fitness of this string. This kind of selection incorporates quite a natural approach, for a string whose fitness is twice as high as the fitness of some other string expects to generate twice as many copies in the mating pool than the other string. As fitness in natural systems is often interpreted as the ability to survive and multiply, this feature seems to be plausible.

On the other hand, this kind of selection operator may lead to severe technical problems. Think for example of a situation, where one string in the population has a comparative high fitness, but is not optimal or near optimal. Let us say the fitness of this string is ten times higher than the average fitness. Proportional selection spreads this string very fast in the population, since the expected number of this string increases by a factor of ten each generation. Therefore, it could easily happen that the population after a few generations consists entirely of these strings. In this case the GA will not evolve further and the optimum will not be found. This phenomenon is called premature convergence and is one of the most imminent problems of the use of GAs.

Another problem of proportional selection is the fine tuning at the end of the search. Consider a case where the average fitness of the population is 10 and the highest possible fitness value is 11. A string encoding the optimal solution expects to generate 1.1 offspring, but as the number of offspring has to be an integer, this string will in most cases be drawn exactly once to send a copy to the mating pool. Considering the disruptive effect of the other genetic operators (they will be described in sections 3.3.2 – 3.3.3), it is highly unlikely, that this optimal solution will take over the whole population. The technical term for this effect is slow finishing .

Both problems, premature convergence and slow finishing, may be avoided if we use scaled fitness values instead of the original fitness values in the selection process. The most popular kind of scaling is the linear scaling, where the scaled fitness value $\bar{f}(i)$ of a string i is given by

$$\bar{f}(i) = \max[af(i) + b, 0]$$

The parameters a and b are calculated for each generation in order to satisfy the following two equalities:

$$\bar{f}^{av} = f^{av} \quad \text{and} \quad \bar{f}^{max} = c\, f^{av},$$

where f^{av} is the average, $f^{\max}$ is the maximal fitness in the population and c is a parameter given by the user to determine the selection pressure. A typical value of c would be 1.5, which means the string with the highest fitness in the population expects 1.5 offspring regardless of the difference between maximal and average payoff.

Another way to circumvent the problems of premature convergence or slow finishing simply is the use of different selection operators. For example, fitness rank selection sorts the individuals according to their raw fitness, and afterwards reproductive fitness values are assigned according to the ranking. The problem of this selection operator is that the raw fitness values have no direct impact on the number of offspring. The string with the highest fitness may be much superior to the rest of the population or it may be just above average; in any case, it will expect the same number of offspring. A special case of this kind of selection is the truncation selection, where the fittest tr percent of the population are selected into the mating pool. The parameter tr is called the truncation threshold. As in this case the size of the mating pool is smaller than the size of the population, the strings in the mating pool have to be mated several times to generate enough offspring.

In many applications, tournament selection has proved to yield superior results than simple proportional selection. In the simplest form, the so-called binary selection two strings are selected randomly from the current population but only the one with the higher fitness value is inserted into the mating pool. This procedure is repeated until the mating pool is full. If larger tournaments than that between two strings are carried out the selection pressure gets higher, because the probability to be selected to the mating pool decreases for individuals with a low fitness, but increases for very fit strings.

Regardless of which selection technique is used the selection operator produces an intermediate population, the mating pool, which consists only of strings that are members of the current population. The following two operators, crossover and mutation, are applied to this mating pool in order to generate the new population.

Parents	0 1 | 1 1 0 1 0 | 0 1 1	$\longrightarrow$	0 1 | 0 1 1 1 0 | 1 1 0	Offspring

Fig. 3.2. One-point crossover with crossover point 2

3.3.2 Crossover

The crossover operator is the key operator to generate new individuals in the population. Inspired by the example of nature, crossover is intended to join the genetic material of strings with a high fitness in order to produce even better individuals. The mating pool is split into $n/2$ pairs of strings and the following operator is applied to each pair with the crossover probability χ. The value of χ is in general larger than 0.6, and often $\chi = 1$ is used. With probability $1-\chi$ no changes are made to both strings, but with probability χ genetic material is exchanged between the two parents. To put this formally, I call a stochastic operator $c : \Omega \times \Omega \to \Omega \times \Omega$ a crossover operator, if for all parents $k_1, k_2 \in \Omega$ with probability one $\{c_1(k_1, k_2)(i), c_2(k_1, k_2)(i)\} = \{k_1(i), k_2(i)\}$ for all bit positions $i = 1, \ldots l$.

In the simplest case of one-point crossover, one crossover point is drawn randomly between 1 and $l-1$. Afterwards the values of the bits to the right of the crossover point are swapped between the two parents. Figure 3.2 shows this procedure for $l = 5$ and the crossover point 2.

Apart from one-point crossover, a number of other crossover techniques were introduced in the GA literature. Some researchers use two-point or even multi-point crossover, where a certain number of crossover points are chosen, and the genetic material is swapped in-between every two of these points. This technique is especially useful if the length of the string is relatively large.

Another quite important crossover variant is the uniform crossover. If this operator is used, first a crossover mask in the form of a binary string is randomly generated. Afterwards the first offspring copies the bit values of the first parent string in all positions where there is a 1 in the crossover mask. Wherever the bit value in the crossover mask is 0, the value from the second parent is copied. The second offspring uses the opposite rule to the first offspring. An example for this kind of crossover is given in figure 3.3. Typically, the crossover mask is generated by a process that assigns the same probability to 0 and to 1, which means that you could just flip a coin for any string position. Whenever "head" appears, the first offspring copies the value from the first parent, when "number" appears he adopts the value of the second parent, and the second offspring will always do the opposite.

Various special crossover operators were developed to generate improved results for special problems. The PMX or OMX operator used in algorithms dealing with the travelling salesman problem are examples of such operators

Crossover mask: 1 0 0 1 0

Parents	0 1 1 1 0	$\longrightarrow$	0 0 0 1 1	Offspring
	1 0 0 1 1		1 1 1 1 0	

Fig. 3.3. Uniform crossover with a given crossover mask

(see Michalewicz [120]). These operators will, however, often work only in one special application, and make merely little sense, if applied to other problems.

Why the crossover operator should increase the performance of the GA will get apparent in section 3.5, where Holland's Schema Theorem and the building block hypothesis are discussed.

3.3.3 Mutation

The mutation operator should allow the GA to find solutions, which contain bit values that are non-existent in the initial population. The parameter governing this operator is called *mutation probability* and will be denoted by μ. After the application of crossover each bit in any string is inverted (set to 1 if it was 0, and set to 0 if it was 1) with probability μ. Mutation probabilities are in general of order 10^{-3}. Whereas the selection operator reduces the diversity in the population, the mutation operator increases it again. The higher the mutation probability, the smaller is the danger of premature convergence. A high mutation probability however transforms a GA to a pure random search algorithm, which is of course not the intention of this algorithm.

3.3.4 Other Operators

Besides the genetic operators presented above, there exist a number of different operators, which are used in different applications. The most widely used non-standard operators are probably inversion and reordering operators, which will however not be described here. Consult Holland [87], Goldberg [74] or Goldberg and Bridges [76] for a detailed discussion of these operators.

In all analytical results derived in this volume only the standard operators are used, while in the simulations in chapter 6 I also use one non-standard operator, which has already been introduced by Arifovic [8] and applied by her to economic systems. It is called election operator and may be described as follows: After crossover and mutation have been applied to the mating pool, for each pair of offspring the fitness in the current environment is calculated. Afterwards parents and offspring compete in a contest and the two individuals of all four with the highest fitness are transferred to P_{t+1}. The implications of the use of this operator will be discussed in chapter 6, and at this stage I only like to mention this operator has proved to facilitate the learning behavior of the GA in various economic applications.

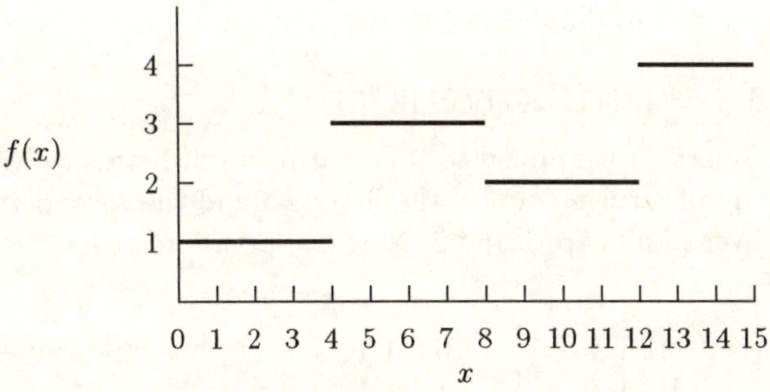

Fig. 3.4. The fitness function to be maximized by the genetic algorithm

3.3.5 An Example

To illustrate the working of a SGA let us explicitly calculate one iteration of the algorithm in a simple example. We use a very simple setup where the population consists only of four binary strings of length 4. Each string encodes an integer between 0 and 15 and the fitness function is given by

$$f(x) = \begin{cases} 1 & 0 \leq x < 4 \\ 3 & 4 \leq x < 8 \\ 2 & 8 \leq x < 12 \\ 4 & 12 \leq x \leq 15 \end{cases},$$

and depicted in figure 3.4. A simple GA with proportional selection, one-point crossover ($\chi = 1$) and mutation ($\mu = 0.001$) is used. Of course, these values for n and l are much too small to be realistic, but this example is only intended to show how a GA actually works.

Let us assume that the initial population P_0 is given by

$$P_0 = \{0010,\ 0110,\ 1011,\ 0111\}.$$

We have

$$f(0010) = 1,\ f(0110) = 3,\ f(1011) = 2,\ f(0111) = 3,$$

which implies that the probabilities to be selected in an arbitrary draw of proportional selection are

$$p(0010) = \frac{1}{9},\ p(0110) = \frac{1}{3},\ p(1011) = \frac{2}{9},\ p(0111) = \frac{1}{3}.$$

As four draws have to be carried out to build a mating pool of size 4 the expected number of copies of a string $i \in P_0$ is given by $4p(i)$, i.e. we expect

4/9 copies of string 0010, 8/9 copies of string 1011 and 4/3 copies of strings 0110 and 0111. Now let us assume that in the first draw 0111 is selected, in the second 1011, in the third 0110, and in the fourth again 0111. This yields the mating pool

$$MP = \{0111,\ 1011,\ 0110,\ 0111\}.$$

Now the crossover operator is applied to this mating pool. Let us assume that the random pairing of strings couples the first two and the second two strings and that crossover points are 1 and 2. Now one-point crossover yields the following transition:

$$\begin{array}{c} \begin{array}{c|ccc} 0 & 1 & 1 & 1 \\ 1 & 0 & 1 & 1 \end{array} \\[6pt] \begin{array}{cc|cc} 0 & 1 & 1 & 0 \\ 0 & 1 & 1 & 1 \end{array} \end{array} \quad \rightarrow \quad \begin{array}{c} 0\,0\,1\,1 \\ 1\,1\,1\,1 \\[6pt] 0\,1\,1\,1 \\ 0\,1\,1\,0 \end{array}$$

Afterwards the mutation operator is applied to these four strings. The mutation probability is $\mu = 0.001$, and there are 16 bits in the population which implies that we may expect only 0.016 mutations. Let us nevertheless assume that one mutation takes place (the bit which is mutated is indicated by a frame):

$$\begin{array}{c} 0\,0\,1\,1 \\ 1\,1\,1\,1 \\[6pt] 0\,1\,1\,1 \\ 0\,1\,\boxed{1}\,0 \end{array} \quad \rightarrow \quad \begin{array}{c} 0\,0\,1\,1 \\ 1\,1\,1\,1 \\[6pt] 0\,1\,1\,1 \\ 0\,1\,0\,0 \end{array}$$

Now we have the new population

$$P_1 = \{0011,\ 1111,\ 0111,\ 0100\}$$

with fitness values

$$f(0011) = 1,\ f(1111) = 4,\ f(0111) = f(0100) = 3$$

and the same procedure starts again. It gets very clear in this example that the transition from P_0 to P_1 is stochastic. However, the stochastic variance in the process gets much smaller if a larger population is used.

3.4 Genetic Algorithms with a Non-Standard Structure

Based on the Schema Theorem, which will be described in section 3.5, it is often suggested that the crossover operator is the main force leading to a good performance of the GA. The mutation operator is only a background operator preventing the GA from loosing some genetic material without any chance of adopting it again. Following this line of argumentation a special kind of genetic algorithm, the so called "gene invariant genetic algorithm" (GIGA) was developed, where no mutation but only crossover like operators are used (see Culberson [41]). On the other hand, Schaffer et al. [151] found in extensive experiments that an algorithm consisting only of selection and mutation can also be a powerful search algorithm. In another paper [152] they stress that an algorithm using only crossover evolves faster than one using only mutation, whereas the mutation-only algorithm in general finds better solutions. However the combination of all three operators seems to be essential for a good performance, especially, in very large spaces and in economic systems where the fitness landscape changes continuously.

Besides the standard structured genetic algorithms described here, there exists a large number of modified algorithms but no overview over these can be provided here. I would just like to mention the variations where not all individuals are replaced in each generation. The simplest case of such a strategy is the elitist strategy where the string with the highest fitness is directly transferred from P_t to P_{t+1} and only the other $n-1$ strings are generated by the application of genetic operators. We will see that this kind of strategy makes it easier to derive analytical statements claiming convergence to optimal solutions in optimization problems.

Finally, it should be pointed out that a very promising line of research is dedicated to self-adaptive Genetic Algorithms where some operators or control parameters adapt while the algorithm is running. It has been observed that such kind of algorithms allow better results especially in large optimization problems than the standard algorithm with inflexible operators and fixed parameter values (see e.g. Hinterding et al. [83], Spears [162] or Srinivas and Patnaik [165]).

3.5 Some Analytical Approaches to Model Genetic Algorithms

The main reason why, genetic algorithms are used in a huge number of optimization problems and other applications, is their large success in these fields and their relative ease of implementation. Many users do not care why the search algorithm they use should find a good solution, but just use the algorithm in order to solve their problem. Since the initial development of genetic algorithms there has been a lot of additional research done in order to improve their performance. However, most of this was based on heuristic arguments or empirical experiments. Mathematical models and analytical results concerning the behavior of genetic algorithms have been rather neglected for a long time, but, recently, the efforts in this field of research have increased dramatically. In this section I shortly outline some of the analytical approaches, which deal with the difficult task to describe genetic learning by a mathematical model that is also analytically tractable. There are a large number of other approaches not being mentioned here, but these are some of the most important concepts.

3.5.1 The Schema Theorem

For a long time the schema theorem has been almost the only mathematical explanation for the success of genetic algorithms. It was developed by Holland [87] and discussed in a large number of publications afterwards. To present the theorem we need to introduce the notion of a *schema*. A schema is the set of all different strings that have certain bit values in various positions. A schema may be written down as a string over the alphabet $\{0, 1, *\}$, where "*" is the "don't care" symbol. The bits having value 0 or 1 are called the specific bits of the schema. A string is called a realization of the schema, if it coincides with the schema in all specific bits. Consider for example the schema s given by

$$s = 0 * 1\,1\,0 * .$$

This schema has the following four realizations:

$$\begin{array}{c} 0\,0\,1\,1\,0\,0 \\ 0\,1\,1\,1\,0\,0 \\ 0\,0\,1\,1\,0\,1 \\ 0\,1\,1\,1\,0\,1 \end{array}$$

It is easy to see that a schema with k "don't care" symbols will have 2^k realizations. On the other hand, every string is a realization of 2^l schemata. This implies that up to $n2^l$ schemata may be present in a population.

The defining length of a schema (denoted by $\delta(s)$) is the distance between the two outmost specific bits, and the order of a schema (denoted by $o(s)$) is the number of specific bits in the schema. For the example above we thus have $o(s) = 4$ and $\delta(s) = 4$.

As already mentioned up to $n2^l$ schemata are present in a population and are transformed by the genetic operators. This means that a huge amount of information is processed in each generation, even if the number of strings in the population is relatively small. This notion of genetic algorithms, which is often seen as the secret of their success, is called implicit parallelism. The schema theorem describes how a certain schema spreads in the population.

Consider a schema s with defining length $\delta(s)$ and order $o(s)$. Let the average fitness of all realizations of this schema in P_t be given by $f(s)$ and denote the average fitness of P_t by $\bar{f}$. The number of realizations of s in P_t will be called $m(s,t)$. Suppose that a simple genetic algorithm with proportional selection and one-point crossover is used. If we calculate the expected number of realizations of s which will be in the mating pool after the selection, we obviously get

$$m'(s,t) = m(s,t)\frac{f(s)}{\bar{f}}.$$

In the next step we calculate the expected number of these realizations which will survive the crossover operator. Consider an arbitrary realization of s. With probability $1 - \chi$ no crossover at all takes place and this realization survives. If crossover takes place the result is still a realization of the schema, whenever the crossover point does not lie between the two outmost specific bits of s. Therefore, the probability that the schema is disrupted by crossover is less than $\chi\frac{\delta(s)}{l-1}$. This implies that the following inequality holds for the expected realizations of s after selection and crossover:

$$m''(s,t) \geq m(s,t)\frac{f(s)}{\bar{f}}\left(1 - \chi\frac{\delta(s)}{l-1}\right).$$

Obviously, a realization of s is still a realization of s after mutation whenever no specific bit of the schema was changed by mutation, which occurs with probability $(1-\mu)^{o(s)}$[1]. Approximating this expression by $(1 - o(s)\mu)$ we get as a lower bound of the realizations of the schema s in P_{t+1}:

$$m(s,t+1) \geq m(s,t)\frac{f(s)}{\bar{f}}\left(1 - \chi\frac{\delta(s)}{l-1} - o(s)\mu\right)$$

This inequality is called the Schema Theorem or fundamental theorem of genetic algorithms.

The implications of the Schema Theorem have been discussed in great detail. The main result is that schemata of low order with a small defining length and an average fitness above the overall average fitness spreads quickly in the population. If their defining length is very small the spreading is almost exponential. These schemata with a relative high fitness but low order and small defining length are called building blocks. The so called building block hypothesis says that in the beginning short building blocks are created, and these building blocks are joined together by crossover to build larger building blocks until a good solution of the problem is reached. It is mainly due to this hypotheses that the crossover operator is seen as the main force in the search and the mutation only as a background operator.

Although the Schema Theorem had and still has a large impact on the research on genetic algorithms, some criticisms have been made (see e.g. Mühlenbein [127, 128]). The most significant shortcoming of the Schema Theorem seems to be that it regards only the selection operator as a possibility to spread a schema, whereas crossover and mutation are merely seen as disruptive operators. However, crossover and mutation are intended to build new schemata with a higher fitness. Therefore, it seems reasonable to analyze explicitly the potential of these operators to build fit schemata, rather than only their disruptive power. But also the assertion about the spreading of schemata due to selection is not so clear as it seems to be. When we talk about a schema with a high average fitness, it is needed to be careful what we exactly mean. If the average fitness of a schema is understood as the average of all realizations of the schema, the statement that schemata whose average fitness is above average increase in frequency is in general not true.

[1] Recall that μ is the mutation probability.

The Schema Theorem says only that a schema will spread if the average of the fitness values of the realizations of this schema in the current population is above average. However, this implies that the statement "schemata whose average fitness are higher than the mean fitness value spread fast in the population", which is the foundation of the building block hypothesis, is only true if the size of the population is sufficiently large.

Besides these we have to face another problem, if we think of economic systems as such rather than of optimization problems. In economic systems the average payoff of a schema depends on the state of the population, and it changes therefore as the population changes. Thus, there exist no building blocks because the schemata with a high fitness will change during the run. Considering all these problems the Schema Theorem does not seem to be a very comprehensive description of the behavior of a genetic algorithm, in particular, if we think of applications in economic systems.

3.5.2 The Quantitative Genetics Approach

The results shortly reviewed in this sub-section were obtained by Heinz Mühlenbein and several of his co-workers using techniques developed in quantitative genetics. Contrary to the approach of the Schema Theorem, here also the positive effect of all genetic operators is taken into account. The analysis is done by means of the response to selection equation, which is used in quantitative genetics. However, some parts of the theory have to be adapted, for quantitative genetics deals with diploids, whereas the strings in a genetic algorithm are in general haploids. Nevertheless, the general approach remains the same.

The general approach might be described as follows. Denote by $r(t+1)$ the response to selection which is defined as

$$r(t+1) = \bar{f}(t+1) - \bar{f}(t).$$

The amount of selection is measured by the selection differential, denoted by $s(t)$ which is given by the difference between the average fitness of the strings in the mating pool and the average fitness in P_t. The prediction of the response to selection is now given by

$$r(t+1) = b(t)s(t),$$

where $b(t)$ is called the realized heritability. . Defining further

$$I(t) = \frac{s(t)}{\sigma(P_t)}$$

as the selection intensity, where $\sigma(P_t)$ is the standard deviation of the fitness values in P_t, we get

$$r(t+1) = Ib(t)\sigma(P_t).$$

Thus, in order to predict the future development of the average fitness, $b(t)$ and $\sigma(P_t)$ have to be estimated. This is in general a very difficult problem and I refer the reader to Mühlenbein [129] and the papers cited therein for an extensive analysis.

Here, I just like to indicate in a very simple framework what kind of insights can be obtained using this approach (see Mühlenbein [128]). Assume that the fitness is determined by the so-called one-max function, where the fitness of a string coincides with the number of 1s in the string. Further consider a genetic algorithm consisting only of proportional selection and crossover. Under these assumptions the selection differential is approximately given by

$$s(t) = \frac{\sigma^2(P_t)}{\bar{f}(t)}.$$

It is easy to see that the average fitness before and after crossover will be the same if the one-max function is used as a fitness function. This implies that $b(t) = 1$ holds. In order to calculate $\sigma^2(P_t)$ let us assume that the number of 1s in a string in P_t follows a binomial distribution with parameters l and $p(t)$. Mühlenbein and Schlierkamp-Voosen [130] show by simulations that the variance in the population will be slightly smaller than the one predicted by the binomial distribution, but the binomial distribution is still a rather good approximation if uniform crossover is used. Under this assumption we get

$$\sigma^2(P_t) = lp(t)(1 - p(t))$$

and

$$\bar{f}(t) = lp(t).$$

This yields

$$r(t+1) = 1 \cdot s(t) = \frac{lp(t)(1 - p(t))}{lp(t)} = 1 - p(t).$$

On the other hand, by definition

$$r(t+1) = \bar{f}(t+1) - \bar{f}(t) = l(p(t+1) - p(t))$$

holds. Equating these two expressions for $r(t+1)$ yields a difference equation, whose solution $p(t)$ can be calculated. By setting $p(t) = 1 - \epsilon$ the expected number of generations needed to reach a state of the population where the fraction of 1s is $1 - \epsilon$ can be calculated. The calculation is straightforward and yields

$$T_{1-\epsilon} \approx l \ln \frac{1 - p(0)}{\epsilon}.$$

This implies, that a genetic algorithm with infinite population and without mutations will reach a state where every bit in the population has its optimal value, only after infinitely many generations, even if a fitness function as simple as the one-max function is used. This seems to be rather disillusioning, but we have to be aware of the fact that we are not interested that all strings in the population are optimal, but we will in general be content even if only one string is almost optimal.

Mühlenbein considers also truncation selection instead of proportional selection and derives by an analogous argumentation as above the following results. If truncation selection and uniform crossover are used and the fitness function is given by the one-max function, the response to selection is given by

$$r(t+1) = I\sqrt{lp(t)(1-p(t))},$$

where again I is the selection intensity. The higher the fraction of strings, which are selected into the mating pool, the lower is I. The expected time till convergence to the optimum can be calculated as

$$T_1 = \left(\frac{\pi}{2} - \arcsin(2p(0)-1)\right)\frac{\sqrt{l}}{I}.$$

Comparison of the terms of convergence time for proportional selection and truncation selection shows that truncation selection is much more effective. A genetic algorithm incorporating truncation selection converges in finite time. The convergence time increases with the length of the strings, but the increase is convex and not linear as with proportional selection. Note that these results only hold only for large population sizes and if no mutation is used. Also, both estimations for the convergence time would increase if a smaller, more accurate estimation of the variance in P_t would be used.

Basically, also mutations can be dealt with in this framework. Rewrite the response to selection with small mutations as

$$r(t+1) = s(t) + s_t\iota - f_t\rho,$$

where s_t is the probability that a mutation increases the fitness of the string, ι is the average increase of the fitness, given that the mutation was successful, f_t is the probability that a mutation decreases the fitness and ρ is the average fitness decrease of an unsuccessful mutation. Mühlenbein and Schlierkamp-Voosen [130] obtain from numerical experiments as well as some simplifications in the analytical terms the following *empirical law:* Given a GA with truncation selection with threshold $tr = 0.5$, and mutations with probability $\mu = \frac{1}{l}$, the response to selection can be approximated by

$$r(t+1) = 2 - 2p(t),$$

and the time until $p(t) = 1 - \epsilon$ is reached is given by

$$T_{1-\epsilon} \approx \frac{n}{2}\ln\frac{1-p(0)}{\epsilon}.$$

This rudimentary sketch of the quantitative genetics approach should indicate how the response to selection equation may be used to estimate the improvement of the average fitness in the population and the convergence time of a GA. We have also seen that this is a suitable tool to judge the efficiency of different GA structures and operators. The main advantage of this approach is perhaps that by operating only with mean values the analytical expressions are quite simple, at least if some assumption concerning the fitness function are made. However, the price which has to be paid for this analytical simplicity is that the model is not complete. The only aspect of the evolving population which is described by the results in this section is the average population fitness. The actual state of the population at some time t can not be reconstructed from these results. This can be a serious problem, especially in economic applications, where the values of the strings – which often encode the strategies of the different agents – might be of higher interest than their fitness values. Also, the simple case presented here is misleading in the sense that it disguises the difficulties in determining the realized heritability in more realistic cases where no such restricting assumptions about the fitness function are made. The analysis will be even more difficult when the fitness of a string depends on the value of all other bits in the whole population. However from chapter 4 onwards we shall always face such situations.

3.5.3 Markov Chain Models

Confronted with the problem to build a model which describes the whole evolving behavior of a genetic algorithm, we have to realize the following facts. Every aspect of the learning behavior of a GA can only be modeled if the representation of the population in the model allows us to reconstruct the exact state of the population in a unique way. Representing a population by its average or maximal fitness is therefore not suited to building an exact model of a GA. Furthermore, it is easy to realize that the state of the population at time $t+1$ depends only on the state at time t. If we know the state of P_t we will get no further information about the state of P_{t+1} from any previous states of the population. As the operators that create P_{t+1} out of P_t depend on several chance events, like the drawing of one certain string to the mating pool or the signal to mutate one certain bit, the dependence of the state of P_{t+1} on the state of P_t is stochastic. These considerations imply that an exact description of the behavior of the GA has to use a Markov chain, where the states of the chain are the possible states of the population [2].

The first work using Markov chains to model the GA is a paper of Goldberg and Segrest [77]. They consider a highly simplified case where each string consists of only one binary bit and only selection and mutation are applied. In this case it is obvious that a state of the population may be described simply by the number of strings in P_t whose single bit has value one. We

[2] See e.g. Adke and Manjunath [1] or Doob [55] for an introduction to the theory of Markov chains and Appendix A.1 for the results used in the following analysis.

have therefore $n+1$ possible states of the population [3]. Goldberg and Segrest first analyze the so-called genetic drift. They assume that both strings, with bit values 0 and 1, receive the same payoff. Nevertheless, the population converges after some time to a uniform population consisting only of one kind of strings. In the simple case considered here it is straightforward to calculate the transition matrix of the corresponding Markov chain in both cases with and without mutation. However, they do not derive any closed analytical expression for the convergence time, but present plots where the convergence time is plotted against the population size for different mutation probabilities. These plots show that the convergence time increases linearly with n for $\mu = 0$, but convex if the mutation probability is positive. As a further step they consider the case where the fitness of the two possible strings is not the same. Let r denote the ratio of the fitness of the string 1 to the fitness of the string 0. In this situation the transition matrix can be calculated explicitly too, but no closed term for the convergence time is given. Numerical calculations for different parameter constellations show that the convergence time decreases for increasing r approximately like $(\ln r)^{-1}$.

Another approach to model the GA with a Markov chain was taken by Rudolph [144]. He no longer makes such simplifying assumptions as Goldberg and Segrest, but deals with a standard 3-operator simple genetic algorithm which he calls the canonical genetic algorithm (CGA). Rudolph argues that a state of the population is given by a binary vector of length nl[4], where the first l bits describe the first string in the population, the second l bits describe the second string and so on. The state space of the process is therefore given by $\{0,1\}^{nl}$. He assumes that proportional selection is used, but does not make any assumption about the crossover operator, for the kind of crossover does not affect his analysis. The three genetic operators transforming P_t into P_{t+1} are described by three transition matrices C, M and S. The complete transition matrix of the process is therefore given by $P = CMS$ (note that Rudolph assumes that crossover is executed first, then mutation and last selection). Simple arguments show that, for positive μ, M is positive, and S has strictly positive diagonal elements. Considering these facts and also the fact that every transition matrix has to be a stochastic matrix (i.e., all elements are nonnegative and the sum of all elements in a row equals 1), implies directly that P is positive. This means, that the Markov chain is irreducible and aperiodic[5], and there exists a unique stationary distribution whose support is the whole state space. From this standard result of Markov theory he concludes, that the maximal fitness within the population does not converge to the optimal value, if convergence is understood in the mathematical sense. The argument is quite simple as mutations will drive P_t out of the states that include the optimal string ever and ever again. On

[3] Recall that n denotes the size of the population.
[4] Vectors are assumed to be column vectors throughout the text.
[5] See Appendix A.1 for the definitions of these properties.

the other hand, Rudolph shows, that, if a kind of elitist strategy is used (the best string in the population is always stored, but in contrast to the original elitist strategy it is not used to create offspring), convergence to the optimal solution is achieved with probability one. The reasons for this fact may be explained without any analytical computations: Obviously it is not possible that the best string in P_{t+1} has a smaller fitness than the best string in P_t(this follows from the fact that the fittest string is transferred to the next generation without change). On the other hand, if the fittest string in P_t is not the optimal solution, there is a positive probability, that the fittest string in P_{t+1} has a higher fitness than the best string in P_t (for $\mu > 0$). This means, however, that every state of the population where the string with the highest fitness is not optimal is a transient state of the chain, and the only closed irreducible set of the chain consists of all states which contain at least one optimal string. From the theory of Markov chains we know, that in such a case the support of the stationary distribution contains only the states consisted in this set, which means that the GA finds the optimal solution with probability one, and never attains a state where this optimal string is no element of the population afterwards. Therefore, the maximal fitness value in $\{P_t\}$ converges to the optimal value for $t \to \infty$.

The results of Rudolph confirm the empirical observations which will be made by any researcher implementing a GA. A simple GA with positive mutation probability never settles down at some state, but keeps oscillating due to the effects of mutation. The fact however that the GA will eventually find the optimal string and never forget it, if the best string of a generation is always directly transferred to the next, is valid not only for GAs, but also hold for any random search algorithm that is basically irreducible.

Looking at the representation of the states of the population in this model, it seems that the state space is larger than what is necessary. As every element of $\{0,1\}^{nl}$ is seen as a different state, two populations which contain exactly the same strings, are in different states, if the strings are not written down in the same succession. In most instances the only relevant information characterizing the state of a population is how often certain strings are contained in the population, but not the succession in which they appear. In the analysis of GAs carried out in the next chapter therefore I do not use the representation chosen by Rudolph, but another model which was developed by Nix and Vose [133]. This model is presented in some detail here, since it is of great importance in order to understand the results of the next chapter.

The model of Nix and Vose is also an exact description of the behavior of the GA. In contrast to the SGA described above they assume that, after selection and crossover have been applied to a selected pair of parents, only one of the two offspring is inserted into P_{t+1}. Which offspring is transferred to the next population is determined randomly, where both offspring have the same probability to be selected. They use a simpler representation of the population states by taking into account that the succession of strings in the population is of no relevance and representing a state of P_t by a vector of dimension $r := 2^l$. To understand this representation we have to realize that any binary string k in $\Omega := \{0,1\}^l$ can be mapped uniquely to the integer given by $\sum_{i=1}^{l} k(i)2^{i-1}$, where $k(i)$ is the value of the i-th bit of k. The integer representing the binary string k will also be called k. Whether I mean by k the binary string or the integer corresponding to the string, will always be clear from the context. If ϕ denotes a state of the population P_t, $n\phi_k$ is the number of strings k in P_t. This means that the k-th component of the string ϕ gives the relative frequency of the string k in the population. The set of all possible states is given by $S := \{\phi \in \mathbb{R}^r | n\phi_k \in \mathbb{N}\ \forall k = 0, \ldots, r-1, \sum_{k=0}^{r-1} \phi_k = 1\}$. To calculate the cardinality of S consider the following different representation of a state of the population: beginning from the left, write for every string 0 in the population a 1, write right of these 1s a 0 and then for every string 1 in the population a 1, then again a 0, for every string 2 a 1 and so on. It is easy to see that in the end the row will consist of $n + r - 1$ digits, that any possible state of the population may be represented like this, and that every such string with n 1s and $(r-1)$ 0s represents exactly one possible state of the population. Thus, the number of possible states equals the number of strings with length $n + r - 1$, where n 1s and $(r-1)$ 0s occur, which is given by

$$|S| = \binom{n+r-1}{r-1}.$$

To show that the number of states is huge even in cases of rather small populations of short strings, I just like to mention that for $l = 12$ and $n = 100$ the cardinality of S is approximately $|S| \approx 10^{200}$. This huge state space is one reason why the analysis of the Markov chain describing a GA is so difficult.

The space S is not only the set of possible states of P, but also the state space of the Markov chain modelling the GA. The next step is to calculate the transition matrix of the Markov chain. Let $p_k(\phi)$ denote the probability that an arbitrary offspring of a population in state ϕ is the string k. Nix and Vose [133] show that the state of the new population is stochastic following a multinomial distribution. The probability that the population is in state ϕ' at time $t + 1$, when it has been in state ϕ at time t, is given by:

$$q_{\phi\phi'} = n! \prod_{k=0}^{r-1} \frac{p_k(\phi)^{n\phi'_k}}{(n\phi'_k)!}. \tag{3.1}$$

The transition matrix of the Markov chain is $Q = [q_{\phi\phi'}]_{\phi,\phi' \in S}$. As the transition probabilities are independent of time, we have a Markov chain with stationary transition probabilities describing the behavior of the genetic algorithm. Using the law of large numbers they prove that the state of the population at time $t + 1$ will be near the expected state at time $t + 1$ if the size of the population is sufficiently large. They reason further that the expected number of strings k in P_{t+1} is due to (3.1) given by $np_k(\phi)$, and the expected frequency of k is therefore $p_k(\phi)$. In their further work Nix and Vose [133] analyze the properties of the expected trajectory of the process. They define the operator $\mathcal{G} : S \to S$ as $\mathcal{G}(\phi) := p(\phi)$ and show that for any $k > 0, \epsilon > 0$ and $\gamma < 1$, there is an $N > 0$ such that $||\Phi_t - \mathcal{G}^t(\Phi_0)|| < \epsilon$ with probability at least γ, for all $n > N$, where $\{\Phi_t\}$ is the trajectory of the Markov chain with transition matrix Q [6]. This implies that for large population sizes the trajectory described by $\mathcal{G}$ is with high probability a good approximation of the real trajectory of the process. The second important result of the analysis of Nix and Vose is the following: Let π^n be the unique stationary distribution of the Markov chain with population size n, let π^* be a weak accumulation point of the π^n for $n \to \infty$, and let $\mathcal{F}$ be the set of fixed points of $\mathcal{G}$. Then $\lim_{t\to\infty} \mathcal{G}^t(\phi) \in \mathcal{F} \; \forall \phi \in S$ implies $\pi^*(\mathcal{F}) = 1$.

These results suggest that the analysis of the operator $\mathcal{G}$ should allow insights into the actual behavior of genetic algorithms with large populations. Some steps towards an analysis of the operator describing the expected trajectory are carried out in a paper of Vose and Liepins [175]. Let M denote the $r \times r$ matrix whose elements are given by $m_{i,j}(0)$ $i, j \in \Omega$, where $m_{ij}(k)$ denotes the probability that an arbitrary offspring of the strings i and j is the string k (this matrix is called the mixing matrix). This implies that the probability that the string 0 results from the reproduction procedure in a population in state ϕ is given by $\phi^T M \phi$ (ϕ^T denotes the transpose of ϕ). Let $\oplus$ be the exclusive or, which implies that $j \oplus k$ is the bitwise difference vector between j and k. Using this notation, symmetry arguments show that

$$m_{k\oplus i, k\oplus j}(k) = m_{i,j}(0)$$

holds, which implies that the probability that a string k is produced by crossover and mutation in state ϕ is given by $(\sigma_k \phi)^T M \sigma_k \phi$, where σ_k is a permutation matrix incorporating a permutation of the basis of the $\mathrm{I\!R}^r$ in a way that the basis vector $e_{j\oplus k}$ is transformed to the basis vector e_j:

$$\sigma_k \left(\phi_0, \ldots, \phi_{r-1}\right)^T = \left(\phi_{k\oplus 0}, \ldots \phi_{k\oplus(r-1)}\right)^T$$

Let $\Delta^r := \{\phi \in \mathrm{I\!R}^r | \phi_k \geq 0 \; \forall k = 0, \ldots r-1, \sum_{k=0}^{r-1} \phi_k = 1\}$ denote the simplex in the $\mathrm{I\!R}^r$. Define the operator $\mathcal{M} : \Delta^r \to \Delta^r$ by setting $\mathcal{M}_k(\phi)$, $k \in \Omega$, $\phi \in \Delta^r$ to the probability that the string k is generated by reproduction in a population in state ϕ. The reasoning above implies

$$\mathcal{M}(\phi) = \left((\sigma_0\phi)^T M(\sigma_0\phi), \ldots, (\sigma_{r-1}\phi)^T M(\sigma_{r-1}\phi)\right)^T . \tag{3.2}$$

Further define a diagonal matrix F, where $F_{kk} = f(k)$ is the fitness of string k. Using proportional selection we get

[6] Notation becomes somehow tricky here. In what follows subscripts of the population state vector Φ may either denote the component or the time considered. In general the meaning will be clear from the context, but to make it even more clear the subscript t is exclusively reserved for indicating the time period.

$$\mathcal{G}(\phi) = \mathcal{M}\left(\frac{F\phi}{|F\phi|}\right). \tag{3.3}$$

$\mathcal{G}$ is a nonlinear operator from Δ^r to Δ^r that describes the expected trajectory of the Markov chain describing the GA. The linearization of the mixing operator $\mathcal{M}$ can be calculated as

$$\mathcal{D}_M(\phi) = 2\sum_{k=0}^{r-1} \sigma_k^{-1} M^* \sigma_k \phi_k, \tag{3.4}$$

with $M^* = [m_{i \oplus j,i}]_{i,j \in \Omega}$. Using this linearization, Vose and Liepins derive the result that for positive μ a fixed point ϕ of $\mathcal{G}$ is asymptotically stable with respect to the dynamical system created by $\mathcal{G}$, whenever the second largest eigenvalue of M^* is smaller than $\frac{1}{2}$. They further conjecture that for one point crossover and $0 < \mu < 0.5$ the second largest eigenvector of M^* is given by $\frac{1}{2} - \mu$, which implies that every fixed point of $\mathcal{G}$ will be asymptotically stable, as long as all strings have the same fitness. This conjecture was latter proven by Koehler [100]. A thorough analysis of the stability of vertices in a standard GA was carried out in Vose and Wright [177] and extended to GAs with strings over an alphabet of higher cardinality than two in Koehler et al. [101]. In both cases they compute the spectrum of the linearization of the dynamical system defined above. I will point out the close connection between their results and the ones obtained for GAs with state dependent fitness functions in chapter 4. Vose and Wright [176] used this approach to show that for linear fitness functions the expected population fitness increases from period to period if the mutation rate is sufficiently small.

At the end of this section I would like to mention some other Markov chain approaches, which can not be described here in detail, although they are of great interest. De Jong et al. [54] use the Nix and Vose Markov model to numerically calculate the transition matrix of the chain, the expected population states for transient periods and expected waiting times for several choices of GA types, parameters and fitness functions. Their analysis sheds new light on the question which functions are hard for GAs and which operators are useful under which circumstances. Davis and Principe [43] try to extrapolate the theory of simulated annealing to GAs and analyze non-stationary chains with decreasing mutation probabilities. Eiben, Aarts and van Hee [57] develop a general theory that includes genetic algorithms and simulated annealing using a Markov approach. Suzuki [163, 164] considers genetic algorithms with an elitist strategy and derives convergence results similar to those of Rudolph. Hartl [80] deals with a variant of a GA which combines the advantages of genetic algorithms and simulated annealing, and shows that this process will under weak assumptions find an optimal solution. A similar result for a different type of mutation operator is derived in Geyer-Schulz [71].

3.6 Do Genetic Algorithms Describe Adaptive Learning?

As already stated above, a genetic algorithm is a simple model of the evolution of natural species. The genetic operators model the dying off of genotypes, whose carriers are not able to generate enough offspring, the creation of new genetic material during the mating, and the changing of single genes due to mutations. However, if we analyze the evolving behavior of a population constituting an economic system, such an interpretation is obviously not very useful. First of all economic learning happens very fast compared to natural evolution. An evolutionary process needs million of years to adapt to its environment, but, of course, an economic agent reacts to changes in the environment within a tiny part of this time. Second, learning in an economic context is not based on the change of the genetic material of the agents. Therefore, we have to face the question, whether the procedure described in this chapter is only a technical algorithm inspired by natural evolution, or whether we may regard GAs also as a suitable model of learning behavior within societies. Basically there are two possible ways to justify the use of a modelling technique to describe adaptive learning processes. Either the assumptions the model rests on are in good accordance with principels of actual decision making or the results are a good match of actual behavior. In the rest of this chapter I will try to discuss in how far these two criteria are met if Genetic Algorithms are used as a learning model. I discuss the second of the two questions first and present in the next subsection several studies where results of genetic learning have been compared with human learning behavior.

3.6.1 Genetic Algorithms and Empirical Evidence

We might distinguish between different types of analyses which have been used in order to evaluate in how far GA results match empirical evidence in economic models. First, it has to be clear what to compare. One might either look at the *individual behavior* induced by the artificial agents' strategies which emerged and compare it to human behavior in similar circumstances or compare *aggregate variables* like prices, aggregate output, exchange rates etc. The underlying question here is, do we intend to build a model of the single agent or are we mainly interested in the overall market behavior. The second distinction concerns the question how to obtain the empirical evidence which is compared to the behavior of the AI agents. One possibility is to perform *laboratory experiments* and expose human agents to exactly the same framework the artificial agents developed in. The second possibility is of course to use data from *real markets*. However, since in general the economic model used cannot take into account all aspects influencing behavior on real markets, often not the actual data is used as a benchmark here but rather 'stylized facts' which should describe the basic observations which can be inferred from the data and are relevant for the model to be analyzed. In what follows I will shortly review some of the applications where GA learning has been compared to empirical evidence. We will see that the different approaches pointed out above have indeed been used in different applications.

In most cases the focus is on the comparision of aggregate behavior but there are execeptions where individual behavior has been analyzed. A very good example is a paper by Midgley et al. [121] where they consider the pricing decisions of different brands in a regional U.S. coffee market. A GA is used to describe the decision of a firm in response to last periods actions of all competitors. In order to reach a tractable length of strings they have to make several simplifications like taking into account only the three major players in the market and allowing only four different price levels. A population of size 25 is considered and the strings are evaluated by playing 50 round games with all possible combinations of the other brands where single-period demands are determined by a one-shot planning simulator. At the end of the simulation there are 25 strategies representing so to speak 25 artificially bred brand managers. Midgley et al. insert these artificial agents into a real life environment by letting them play against the recorded actual decisions of the other two brands during a year. As well the actions as the profits of these artificial managers have afterwards been compared to those of the real brand managers. It turned out that the best string in the population could gain more than twice the profit actually earned by this brand during the year and none of the artificially bred managers performed significantly worse than the brand actually did. An interesting observation of Midgley et al. is also that the behavior of the worst strings in the population looked very similar to that recorded in the real market, whereas the best strings in the population reduced the prices much more often than the actual brand managers did. A possible explnantion of the authors for this phenomenon is that actual brand managers cannot react to competition on a week-by-week basis but have to plan promotions far ahaed. Thus institutional constraints may be the reason for this rather big performance gap between the human and artificially intelligent brand manager. Unfortunately they do not extend their analysis to the case where the stylized institutional constraints are also binding for the artificially intelligent brand managers.

A different approach for testing GA learning against empirical data is taken by Arifovic in several contributions. In several standard models she either uses data from laboratory experiments carried out by other researchers or performs own experiments in order to obtain series of aggregate variables which stem from human learning behavior. The first paper which was published where this approach is used deals with the standard cobweb model [8]. I have shortly described the setup in section 2.4.3 and we will deal with this model in more detail in chapter 6. Wellford [182] published in 1989 results of laboratory experiments performed in a simple cobweb framework. In the incomplete information setup the agents only have information about the last-period price of the good and their own earned profits when they decide on their output quantity for the current period. This is about the same information available to the artificial agents in the GA environment and therefore Arifovic uses the data from these experiments to evaluate the GA learning behavior. Wellfords experiments broadly speaking showed that the average output quantity always fluctuates around the rational expectations equilibrium and that the amplitude of the fluctuations decrease with time. This holds also for the so-called unstable case where the quantities would diverge if agents used naive expectations and produce the optimal output with respect to these expectations. Arifovic shows in her paper that these two facts – a fluctuating approach towards the equilibrium and convergence also in the unstable case – cannot be reproduced by using standard techniques for the expectation updating. Naive expectations and least square learning rules lead the divergence in the unstable case. Moving average learning yields convergence in the stable and unstable case but convergence is smooth rather than fluctuating. On the other hand, the average output quantity in the GA simulations shows fluctuating convergence towards the rational expectations equilibrium in both the stable and and the unstable case. Also it was shown that – like with human agents – the price variance in the unstable case is larger than in the stable one. So GA learning yields qualitatively similar aggregate behavior than a population of human agents. The match is not perfect since the amplitude of the oscillations decreases faster in the GA runs (in particular in the unstable case) but compared to the other approaches mentioned these results are very satisfying.

A similar picture emerges in quite a different framework, namely that of an overlapping generation model of a monetary economy[7] (Arifovic [9]). With constant money supply there exists a unique monetary equilibrium and as well in experiments as in simulations with GAs the inflation rate converges towards this value. The same holds also true for many other learnign rules. More interesting in this context is the case where a fixed real deficit is financed by the government through seignorage. In such a case a low and a high inflation stationary equilibrium exist. Experiments with human agents performed by Marimon and Sunder [116] and Arifovic [7] show that inflation always converges towards the low inflation stationary equilibrium where convergence is not smooth but the time series exhibits decreasing fluctuations. Least square learning does not show these fluctuations and also diverges for certain parameter constellations although the experimantal inflation rate converges. Again, the GA does a better job in reproducing the behavior that was observed in the experiments. Here the inflation rate always converges towards the low inflation stationary equilibrium value and also fluctuations can be observed. However – like in the cobweb case – the fluctuations are more persistent in time series stemming from experiments than in those created by using GAs.

[7] See chapter 6

In Arifovic [10] the analysis is extended to a model with two currencies. The experiments in this framework were performed by Arifovic herself and in the laboratory the exchange rate between the two currencies showed persistent fluctuations. On the other hand, first-period consumption of the individuals converges only in the beginning of the experiment but afterwards stays more or less constant close to the stationary equilibrium level. These properties can also be observed in the trajectories generated by GA learning. Arifovic also performs several statistical tests to establish similarities in the qualitative properties of the the time series. However, in this case no comparision with other learning algorithms is carried out, so we do not know in how far these properties are generic in a sense that they would appear for any sensible learning process.

Also in Arifovic [11] the learning behavior of GAs is compared with evidence gathered in experiments with human agents. Van Huyck et al. [170] used in their laboratory experiments a simple coordination game where n players independently choose a level of effort where the costs increase linearly with the level of effort. The output, on the other hand, is a linear function of the minimum of all effort levels chosen. All states where all players choose the same effort level are Nash equilibria and the state where all choose the highest level of effort is Pareto efficient. The experiments show that with large group sizes (14 players) coordination failures occur and the state where all agents choose the minimal effort level is reached. However, if the number of players is small (2 players) the Pareto efficient equilibrium is the outcome. These effects which may be explained by the fact that due to strategic uncertainty players want to be save and choose their maximin strategies if the group is to large can also be observed in simulations where Genetic Algorithms describe the learning behavior of the agents. In this implementation a string in the population determines the current level of effort of the player in dependence on the minimum effort of his group in the previous iteration. In all runs the players reached an equilibrium but only for small group sizes (2-4 players) play converged to the payoff dominant equilibrium. For group sizes larger than 5 the secure equilibrium where all players choose the minimal effort emerged. Again, experiments with human and artificially intelligent agents represented by a GA yield qualitatively very similar results.

Whereas in the contributions of Arifovic the empirical evidence always stemmed from laboratory experiments with human agents, other researchers have tried to explain stylized facts observed in real world markets by adaptive learning models based on GAs. One example here is the paper by Arthur et al. [13] which has already been described in section 2.4.4. They point out that price volatility and trading volume on financial markets show persistence or autocorrelation and that there is cross-correllation between volatility and volume. These stylized facts can also be observed in the artificial stock market created at the Santa Fe Institute. The authors compare the corresponding numbers for their complex regime with those of actual IBM stock data and show that they qualitatively look similar although there is no exact match.

Lettau [109], on the other hand considers mutual fund flows. He groups the funds into four categories with different degrees of risk and uses data from the Investment Company Institute to analyze the flows between these four categories. The following three stylized facts were obtained in this analysis: (i) flows into mutual funds are positively correlated with returns, (ii) flows are more sensitive to negative returns than to positive ones, and (iii) evidence is stronger for riskier mutual funds. These facts cannot be explained in the framework of rational agents but also simulations with a simple GA (whith an exponentially decreasing mutation rate) did not exhibit these properties. However, changing the algorithm slightly such that a small number of randomly selected strings in the population is every period replaced by new strings with random values, enabled Lettau to obtain time series of fund flows which have the desired properties. Again, this is only meant in a qualitative sense, since for example the coefficient describing the dependence of flows on returns is significanlty bigger in the GA simulations than in the real market data.

These examples show that GA based models were able to generate qualitatively correct descriptions of aggregate and individual behavior in laboratory experiments and real markets. Whether we should be satisfied with a qualitatively correct description depends on the purpose of the analysis but to expect from a model of adaptive learning in economics to be more than qualitatively correct seems to be rather unrealistic. Of course this selection of examples is biased and there may be other applications where GAs did a very bad job in explaining empirical data, which – exactly for this reason – have not been published. Also, it might be that in some cases the models or the algorithm used have been specially tuned in order to obtain satisfying results. A more systematic comparision of GA learning with other approaches frequently used which is based on rigorous statistical analysis and a wide field of applications would be very helpful in this respect but to my knowledge no such project has been started yet. Nevertheless, the fact that the examples presented above come from completely distinct fields of economics suggests that there is a broad spectrum of economic problems where a GA based learning model can indeed be used as a suitable model of adaptive human behavior. Considering the rather encouraging empirical evidence the question arises whether a GA should only be seen as a black box algorithm which is able to describe empricial data well or whether the actual algorithm can be motivated by some underlying behavioral rules which are plausible in an economic context.

3.6.2 Economic Interpretation of the Algorithm

In a recent publication Goldberg [75] points out that GAs are a computational model of innovation and creativity. He argues that a human being, faced with some new problem, will often bring together several solution features which worked well in different contexts in order to solve the problem. Goldberg shows that these features are also present in a genetic algorithm and concludes that "*Economies, organizations, social systems, political systems, whole societies, seem a little less mysterious after one has spent some time watching GA runs and learning from them. The reasons for this are twofold. First, much of the mystery of such systems emanates from their innovative nature, and GAs replace the mystery shrouding innovation with a healthy dosage of mechanism. Many of the difficulties in the social sciences come from the lack of a computational theory of actor innovation, and genetic algorithms increasingly fill that bill quite nicely. Second, population-oriented systems are dominated by what economists call the law of unintended consequences (which is itself largely the result of the innovative capability of the actors) and interacting with GAs provides hands-on experience in understanding what for most people is counterintuitive behavior.*" This quote shows that also scientists who are not a priori interested in economics have realized that GAs are not only an optimization tool but that insights may be obtained if GAs are interpreted as a model of a learning population. Apart from the empirical evidence mentioned above this point made by Goldberg seems to be the main motivation for the use of GAs in Economics. They give us an opportunity to model in a very flexible way the evolving behavior of a population without loosing completely control over the model. But besides this general view a microfoundation of GAs as a model of learning would be very useful for an even better understanding of the aggregate dynamics in an evolving population.

Brenner [35] recently addressed the question whether evolutionary processes describe learning processes. He uses a continuous time dynamical system in form of a selection-mutation-equation to give a general description of evolutionary processes and compares the structure of the dynamics to that of a Variation-Imitation-Decision (VID) model which he develpoed earlier (Brenner [34]) as a general description of a social learning model. A thorough comparison of the analytical terms occuring in both dynamical systems implies that several apsects of the VID model cannot be accounted for in the evolutionary dynamics. Brenner draws the conclusion that using evolutionary algorithms to model social evolution is problematic.

In my opinion the fact that there are aspects of adaptive learning which cannot be accounted for in GA models is no reason to abandon the use of GAs in this field alltogether. However, it is very important to have a clear understanding of what kind of learning behavior is actually described by a GA model. Here, I will try to develop an interpretation of genetic learning which shows what aspects of adatpive learning are present in such a model.

First, we have to decide as how to interpret the most basic components of a GA; namely, the binary strings. It is quite natural to consider every string in the population as the representation of one economic agent. There is also another way to think of the strings in the GA, namely, as different ideas of one individual which fight against each other to get executed (Arifovic [8]). Vriend [179] denotes the case where each string represents one individual as social learning and the case where a whole population represents on decision maker as individual learning. He shows in the framework of a Cournot oligopoly game that these two types of learning may lead to qualitatively different long run outcomes. In this monograph GAs will always be considered as a model of social learning where each string represents an individual. This is also the underlying assumption in the following interpretation of the operators.

In considering the selection operator, the economic interpretation seems to be quite straightforward. In a population where the payoffs of the actions of every individual are known to all other members of the population, it is very plausible to assume that the individuals with a low payoff will imitate the actions of the more successful agents. As the fitness of a string in an economic model is given by the payoff of the action encoded by the string, this kind of "imitation of the successful" may lead to a spreading of the strings encoding actions with a high payoff and the vanishing of strings encoding very poor strategies. The selection operator may therefore be seen as the modelling of an imitation effect within a population. The importance of imitation effects has been stressed recently in various learning models (e.g. Friedman [65], Vega-Redondo [171, 173], Björnerstedt and Weibull [30], Dawid [45, 47]). Of course, there are several ways to model this phenomenon, but the selection operator incorporates the basic feature that better paying strategies become more prevalent. As Mailath [112] puts it, "Imitation is often an important part of learning; successful behavior tends to be imitated. In addition, successful behavior will be taught." (p. 264). Empirical evidence for the presence of imitation effects in learning is for example provided by Pinlge [137]. The imitation effect accounts for the observed uniformity among the survivors. If one thinks of our individuals in the population as representing firms, one might object to such an approach that firms do not reproduce and do not die. Note, however, that we changed the perspective by identifying the firm by a strategy, which can be e.g. a production method. Now the different production methods are the objects under selective pressure, and these may well be born or die.

Considering the crossover operator, we might interpret the results of this operator as a communication or more general as an information exchange effect. Two members of the population meet and tell each other the plans for their next actions. Some of the individuals adopt a part of the strategy of the other agent and use it to change their own plan. This kind of communication effect is not common in economic literature, but it makes sense to assume that an individual may be influenced by the other members of the population, even if it does not just imitate another individual.

As it is, several criticisms may be brought up against this interpretation. First, it is in some cases difficult to give a sound interpretation of the "meaning" of single bits or substrings. This may crucially depend on the values of their surrounding bits and therefore a substring which is adopted from some other string may have quite different effects in the new string. But, on the one hand, this problem may be circumvent by using an appropriate coding mechanism, and, on the other hand this effect is not so unrealistic. If a firm adopts only one aspect of the strategy of a competing firm, it may well be that this aspect does not fit exactly into its own strategy and has non-anticipated effects. Another objection may be that any exchange of information implies that the information which is transferred to another agent is lost for the donor of the information and that such a scenario is unrealistic. However, in economic systems a string encodes in general a strategy which determines what to do in different circumstances. Thus for any possible state of the surrounding system a unique action is encoded and each agent is characterized by the set of all these actions. In this setup, the information about the success of other actions is not preserved by the single individual but by the whole population.

To make this kind of interpretation of the crossover operator more clear, let us consider Axelrod's simulations in the iterated prisoner's dilemma which were described in section 2.4.2. While looking at figure 2.4 we realize that any bit has a clear meaning, either it gives a subjective estimation of the sequence of play before the first period, or it says how to react to a given sequence of play. If due to crossover the bit at position 13 changes from 0 to 1 this means that this agent was influenced by another agent he met to cooperate instead of defect whenever the previous sequence of play was $DDCCDC$ $(13 = 0 \cdot 2^5 + 0 \cdot 2^4 + 1 \cdot 2^3 + 1 \cdot 2^2 + 0 \cdot 2^1 + 1 \cdot 2^0)$. This example shows quite clearly that crossover may be very well suited to describe the exchange of information within a population.

The distinction in the interpretation of selection and crossover as imitation and communication is somehow fuzzy. Obviously, some kind of communication is often needed in order to allow for successful imitation and in this stylized model it is not made explicit how the information needed for imitation is distributed. But, as pointed out before, the interpretation provided here is on a very abstract level.

The mutation operator incorporates the effect of innovations or mistakes of the agents. Individuals change their actions randomly either by mistake, or because they think the best actions have not been discovered yet. Randomly made mistakes are a feature which is extremely common in economic models. There are even special equilibrium concepts filtering the steady states which are stable with respect to such perturbations (Selten [154]). On the other hand there are some examples where innovations which proved to be extremely useful were made randomly. Thus also the mutation operator has a clear economic interpretation.

Some of the ideas presented above are of course not new in economics. As one of the first, Alchian [4] in his seminal paper explicitly recognized the use of imitation in contrast to optimization for guiding much of economic behavior. He states that, "uncertainty provides an excellent reason for imitation of observed success" and further concludes that, "*Imperfect imitators provide opportunity for innovation, and the survival criterion of the economy determines the successful, possibly because imperfect, imitators.*" As the economic counterparts of genetic heredity, mutations, and natural selection he relates to imitation, innovation and positive profits. Our ideas are also incorporated in evolutionary theories in economics, where "*Evolutionary models in the social domain involve some processes of imperfect (mistake-ridden) learning and discovery, on the one hand, and some selection mechanism, on the other.*" (Dosi and Nelson [56]).

To summarize, we may describe a population whose adaptive behavior is modeled by a genetic algorithm as follows. A population is considered where each individual executes some action and gains a payoff for this action which depends on the actions of all the other individuals. The individuals, either due to limited knowledge or due to limited computational capacity, by and large fail to calculate their optimal action. Instead they adopt an action which has been used by some other individual in the previous period and prefer actions which proved to be successful. They may also be influenced by other individuals in the population and change parts of their strategy due to this influence. And finally the action executed may also be subject to random errors or innovations.

Without doubt there are other models of a population with these features which are more straightforward than that of a GA. The interpretation provided here gives us a clear idea how to expound the effects of the different operators from an economic point of view. This knowledge is helpful to understand the results in the next chapters. This chapter has demonstrated that the appeal of GAs in economics to a large part stems from the fact that they are a flexible tool to describe adaptive learning which in many instances matches empirical data quite well.

4. Genetic Algorithms with a State Dependent Fitness Function

4.1 State Dependency in Economic Systems

In the last chapter I have mainly reviewed existing literature and models, which have proven to be of great importance for the theoretical analysis of genetic algorithms. In this chapter we will deal with a new problem, which has not been dealt with in this literature[1]. The analytical models presented in chapter 3 all assume that the genetic algorithm is used to solve an optimization problem for an exogenously given fitness function. However, this is not the case if we think of an economic system, like a market, where the payoff of a single market member depends crucially on the actions of the rest of the market. The same argument holds also for two agents playing a normal form game, where the payoff of a strategy depends on the opponents' strategy. In such models which are also often called co-evolutionary models two major aspects of the genetic algorithm change compared to its traditional application as an optimization tool. Firstly, the fitness of a single string depends on the state of the whole population. In an optimization problem the fitness values can be written in one r-dimensional vector f, but in economic systems the fitness is in general given by a r-dimensional function $f : S \rightarrow \mathbb{R}^r$, where $f_k(\phi)$ is the fitness of string k, when the whole population is in state $\phi \in S$[2]. Secondly, we are no longer interested in the learning of optimal solutions, but rather in the question whether adaptive learning eventually leads to equilibrium behavior. As the simplest concept of an economic equilibrium we might say that the system is in equilibrium if the current action of every agent is optimal under the assumption, that all other agents behave according to the equilibrium. For later reference a formal definition of an economic equilibrium is given in definition 4.1.1. In what follows, I will often stress the term "economic" to distinguish it from dynamic equilibria.

Definition 4.1.1. *A state $\phi \in S$ corresponds to an equilibrium of an economic system defined by $f(\cdot)$ if and only if*

$$f_k(\phi) \geq f_j(\phi) \ \forall j \in \Omega$$

[1] Parts of this chapter are based on Dawid [44] and Dawid and Hornik [51].

[2] The notation introduced in subsection 3.5.3. is again used in this chapter

for any $k \in \Omega$ with $\phi_k > 0$. The equilibrium is strict if the above inequality is strict for all j with $j \neq k$.

As the names for equilibrium refinements are not coherent in the literature I would like to mention that this definition of a strict equilibrium was adopted from Fudenberg and Tirole [63]. Note that any strict equilibrium state has to be uniform.

In what follows a system with a state dependent fitness function will be abbreviated as an SDF system. There are also non-economic SDF systems, for example, in ecological or biological models, but also every genetic algorithm with scaling is an SDF system. Obviously the scaled fitness of a string depends on the raw fitness of all the other strings, and therefore on the state of the population. All the analytical approaches up to now assume, however, that the fitness of a string depends only on the bit values of this string. If we want to analyze the behavior of any GA with scaling, we should not use the results reported in the last chapter, but rather use results for SDF systems instead. Thus, the theoretical results which will be derived in this chapter are not only important for economic applications of GAs, but also for all applications where GAs with scaling are used. However, the task here is the analysis of economic systems. To my knowledge no mathematical analysis of genetic learning in SDF systems has been carried out yet. Thus we are on new ground here but we may still rely on the techniques and models which were developed for the analysis of the behavior of GAs in optimization problems.

4.2 A Markov Model for Systems with a State Dependent Fitness Function

The Markov model used in the following analysis is an adaptation of the model of Nix and Vose [133], presented in section 3.5.3, to SDF systems. I again assume that the population P consists of n binary strings of length l, denote by Ω the set of all possible binary strings of length l, and by S the set of possible states of the population. A state of the population is again given by a r-dimensional nonnegative vector, describing a frequency distribution over Ω. However, the fitness of a string is now given by a state dependent fitness function $f : S \to \mathbb{R}^r_+$. I use this special representation out of all models introduced in section 3.5 because this kind of representation is the simplest form which allows a complete, exact analysis. As already explained in the last chapter this representation enables us to discriminate between states which contain different strings, but, on the other hand, it does not discriminate between states where the same strings are contained in different order. Furthermore, I assume that a simple genetic algorithm with proportional selection is used in the entire analysis. Since the basic considerations concerning the behavior of the GA do not depend on the actual fitness values of the single strings, it is easy to conclude that the process defined by the series of states of $\{P_t\}$ is again a time homogeneous Markov process, with a transition matrix Q. Unfortunately the claim that the state at time $t+1$, given the state of P_t is ϕ, is multinomially distributed with parameters n and $p(\phi)$ is no longer true in our setup, for here both offspring of every crossover are inserted into the next population. This observation is independent from the fact whether the fitness function is state dependent or not. Consider the following simple example with a state independent fitness function.

Given is a binary population consisting of 2 binary strings of length 2. The set Ω, therefore, contains 4 strings which are denoted by their value in the binary representation, i.e. $\Omega = \{00, 01, 10, 11\}$. Assume further that the state of P_t is given by $\phi = \frac{1}{2}(0,1,1,0)$, which means that there is one string 01 and one string 10 in the population. We apply now the genetic operators proportional selection and uniform crossover to this population. For reasons of simplicity let all strings have the same fitness and let $\chi = 1$. Next we calculate the value of $p_0(\phi)$. The string 00 could only be created by crossover between 01 and 10. The probability that exactly these two strings are selected into the mating pool is given by $\frac{1}{2}$. Given that the mating pool is $\{01, 10\}$, the probability that an arbitrarily chosen offspring of uniform crossover between the two strings is 00 is given by $\frac{1}{4}$. We therefore get $p_0(\phi) = \frac{1}{8}$ and in the same way

$$p_0(\phi) = \frac{1}{8}, \qquad p_1(\phi) = \frac{3}{8}, \qquad p_2(\phi) = \frac{3}{8}, \qquad p_3(\phi) = \frac{1}{8}.$$

And, with $\phi' = \frac{1}{2}(1,1,0,0)$ we get $q_{\phi\phi'} = \frac{3}{32}$ if we insert these values into (3.1). However, the only mating pools which may be generated from the state $\frac{1}{2}(0,1,1,0)$ are $\{01,01\}$,$\{01,10\}$ and $\{10,10\}$. Obviously the offspring of crossover between the two strings in the mating pool can in no case be $(00, 01)$. The correct value of the transition probability therefore has to be $q_{\phi\phi'} = 0$. Mathematically, the events $\{c_1(i,j) = k_1\}$ and $\{c_2(i,j) = k_2\}$ are not independent, which implies that the offspring of every single crossover is not multinomially distributed, and therefore also the distribution of the state of P_{t+1} is not distributed according to (3.1).

Hence, we have to face the fact that we are not able to give a simple characterization of the distribution of ϕ_{t+1}. Nevertheless, we may give a characterization of the limit distribution of the Markov chain described by the transition matrix Q. First, observe that an arbitrary string can be transformed into any other string just by the means of mutations. This means that for $\mu > 0$ the transition matrix Q of the process modelling the behavior of a GA is strictly positive. This means that the chain is irreducible, and using standard results from the theory of Markov chains (see Appendix A.1) we get that the process has a unique stationary distribution, which coincides with the limit distribution of the chain.

The first proposition will show that this limit distribution is concentrated on the uniform states if μ is sufficiently small. A state is said to be uniform if all strings in the population have equal bit values. In my notation, a uniform state consisting only of strings k is represented by the unit vector e_k. The theory of perturbed Markov chains will be used in order to establish this result.

First, I introduce some notation which will be used. For any state $\phi \in S$, a ϕ-graph g is a graph on S with the following properties:

(i) Every state $\phi' \neq \phi \in S$ is the initial point of exactly one arrow.

(ii) There is no arrow for which ϕ is the initial point.

(iii) For any state $\phi' \neq \phi \in S$ there exists a sequence of arrows leading from ϕ' to ϕ.

The set of all ϕ-graphs is denoted by $G(\phi)$. With this notation we get the so called "Markov chain tree theorem" which was first proven by Hill [82] and independently also by Shubert [158] (a proof of the theorem is given in the Appendix).

Theorem 4.2.1. *Consider an irreducible Markov chain on the finite state space S with transition matrix Q, then the unique stationary distribution π is given by $\pi = (\sum_{\phi \in S} \nu_\phi)^{-1} \nu$ with*

$$\nu_\phi = \sum_{g \in G(\phi)} \prod_{(x,y) \in g} q_{xy} \quad \forall \phi \in S.$$

Using this theorem we can prove the following proposition:

Proposition 4.2.1. *Let π be the stationary distribution of the Markov chain describing a GA, then*

(i) $\lim_{\mu \to 0} \pi_\phi = 0 \quad \forall \phi \neq e_k \;\; k = 0, \ldots, r-1.$

(ii) $\lim_{\mu \to 0} \pi_\phi > 0 \quad \forall \phi = e_k \;\; k = 0, \ldots, r-1.$

Proof. Note first that all transition probabilities $q_{\phi\phi'}(\mu)$, $\phi, \phi' \in S$ are polynomials in μ, and therefore also $\nu_\phi(\mu) = \sum_{g \in G(\phi)} \prod_{(x,y) \in g} q_{xy}(\mu)$ are polynomials in μ. Furthermore, it is obvious that all uniform states e_k, $k = 0, \ldots, r-1$, are absorbing for $\mu = 0$. This means that mutations are needed to come from any uniform state to any other state. Therefore, all transition probabilities $q_{e_k,\phi}$ are polynomials in μ, where the lowest power of μ is greater or equal to 1. Let $o(\phi, \phi')$, $o(g) := \sum_{(x,y) \in g} o(x,y)$, and $\omega(\phi) := \min_{g \in G(\phi)} o(g)$ denote the smallest powers of μ in $q_{\phi\phi'}$, in a graph g, and in ν_ϕ, respectively. Now let $\phi \in S$ be any non-uniform state, i.e., there are strings $j, k \in \Omega$, $j \neq k$ with $\phi_j > 0$ and $\phi_k > 0$, and let h be a ϕ-graph with $o(h) = \omega(\phi)$. We will construct an e_k-graph h' with $o(h') < o(h)$, which implies that $\omega(e_k) < \omega(\phi)$.

This graph h' is derived from h by deleting the arrow leaving e_k (call the endpoint of this arrow z) and adding an arrow from ϕ to e_k. By routine verification of conditions (i)–(iii) one can see that h' is an e_k-graph. As obviously

$$q_{e_k z} \prod_{(x,y)\in h'} q_{xy} = q_{\phi e_k} \prod_{(x,y)\in h} q_{xy},$$

we have

$$o(h') = o(h) + o(\phi, e_k) - o(e_k, z).$$

As $\phi_k > 0$, there is a positive probability that the selection process will create a mating pool consisting only of strings k, which implies that no mutations are needed to change the state of the population from ϕ to e_k. Hence, $o(\phi, e_k) = 0$. As, on the other hand, e_k is a uniform state which implies $o(e_k, z) \geq 1$, we infer that $\omega(e_k) \leq o(h') < o(h) = \omega(\phi)$. That is, $\nu_\phi(\mu)$ tends to zero at an order strictly higher than $\nu_{e_k}(\mu)$. Direct application of theorem 4.2.1 now establishes assertion (a) of the proposition.

To prove (b), we have to show that $\omega(e_k) = \omega(e_j)$ for all uniform population states e_k and e_j. Let g be an e_k-graph with $o(g) = \omega(e_k)$ and j any string in Ω with $j \neq k$. Let the graph h be defined by the rule

$$(x, y) \in h \Leftrightarrow (\sigma_{k\oplus j}x, \sigma_{k\oplus j}y) \in g.$$

As $\sigma_{k\oplus j}e_k = e_j$, it is easy to see that h is an element of $G(e_j)$. From the definition of $o(h)$ we get

$$o(h) = \sum_{(x,y)\in h} o(x, y) = \sum_{(x,y)\in g} o(\sigma_{k\oplus j}x, \sigma_{k\oplus j}y).$$

Thus, it has to be shown that the smallest power of μ in q_{xy} equals the smallest power of μ in $q_{\sigma_i x \sigma_i y}$ for all strings $i \in \Omega$ and all states $x, y \in S$. Let $\tilde{q}_{\phi\phi'}$ and $\hat{q}_{\phi\phi'}$ denote the probabilities that the state of the mating pool at time $t+1$ is ϕ' given that P_t is in state ϕ, and that P_{t+1} is in state ϕ' given that the mating pool at time $t+1$ is in state ϕ, respectively. Then

$$q_{xy} = \sum_{\phi\in S} \tilde{q}_{x\phi}\hat{q}_{\phi y}. \tag{4.1}$$

Now take two arbitrary strings j_1 and j_2, apply crossover with some crossover mask to these strings and mutate certain bits. Let us call the offspring k_1 and k_2. It is easy to realize that crossover with the same crossover mask and mutation at the same bits applied to the parents $j_1 \oplus i$ and $j_2 \oplus i$ yields $k_1 \oplus i$ and $k_2 \oplus i$ as offspring. Hence,

$$\hat{q}_{\sigma_i x \sigma_i y} = \hat{q}_{xy}, \quad \forall x, y \in S, \quad \forall i \in \Omega.$$

As $\tilde{q}_{\phi\phi'}$ describes only selection effects, it obviously does not depend on μ. Note that due to the fact that the fitness function is strictly positive, $\tilde{q}_{\phi\phi'}$ is positive if and only if $\mathrm{supp}(\phi') \subseteq \mathrm{supp}(\phi)$. Therefore, the equivalence

$$\tilde{q}_{xy} > 0 \Leftrightarrow \tilde{q}_{\sigma_i x \sigma_i y} > 0 \tag{4.2}$$

holds for every $i \in \Omega$, from which we obtain that

$$\begin{aligned} q_{\sigma_i x \sigma_i y}(\mu) &= \sum\nolimits_{\phi \in S} \tilde{q}_{\sigma_i x \phi} \hat{q}_{\phi \sigma_i y}(\mu) \\ &= \sum\nolimits_{\phi \in S} \tilde{q}_{\sigma_i x \sigma_i \phi} \hat{q}_{\sigma_i \phi \sigma_i y}(\mu) \\ &= \sum\nolimits_{\phi \in S} \tilde{q}_{\sigma_i x \sigma_i \phi} \hat{q}_{\phi y}(\mu). \end{aligned}$$

Comparing this expression with the expression in equation 4.1 we conclude by using (4.2) that μ appears with exactly the same powers in both expressions (the coefficients may of course be different). This implies that $o(g) = o(h)$, from which we immediately infer that $\omega(e_j) = \omega(e_k)$ for all j and k in Ω. Application of theorem 4.2.1 now completes the proof.

At first sight the obtained result seems to be rather trivial. It is easy to see that for $\mu = 0$ the chain is no longer irreducible and that all uniform states are absorbing without mutations. Given this, it seems natural to expect that the limit of the limiting distribution for $\mu \to 0$ must consist of all uniform states. However, this conclusion is in general false. There are various examples of Markov chains modelling processes consisting of a basic dynamic disrupted by mutations, where there exist several absorbing states without mutation, but the support of the limit of the limiting distribution for decreasing mutation probabilities does not contain all these states (see for example the model of Kandori, Mailath and Rob [91]). Therefore, the result of proposition 4.2.1 is not trivial at all.

Proposition 4.2.1 can be seen as an extension of theorem 4 in Rudolph [144], which states that a canonical genetic algorithm with proportional selection and strictly positive mutation probability does not converge to the global optimum. The argument of Rudolph is that, since the chain is irreducible, the GA does not converge to any uniform state, which of course implies that it also does not converge to an optimum. Proposition 4.2.1 states that, although the support of the limit distribution does not consist entirely of uniform states, for small mutation probabilities the population is after some transient period in a uniform state for almost all the time. This result is consistent with empirical observations, which show that a GA typically converges to a state that is basically uniform and only disrupted by single mutations.

On the other hand, several questions remain open in proposition 4.2.1. First of all, it is an ergodic result which means that the characterization given holds only after a long period of time. However, we have no clear result stating how long we have to wait till the process settles down near an uniform state. Furthermore we do not have an estimation which indicates how small the mutation probability has to be to ensure that not more weight than some given δ lies on non-uniform states. And, of course, we do not know how much weight lies on a certain uniform state. What we can deduce from proposition 4.2.1 is that heterogeneous states are structurally unstable . Thus, we will restrict our attention in the rest of the chapter to homogeneous states and will try to answer the question which uniform state has a large limit probability and which has a small one.

To gain some insight into this problem, in the next sections some results are derived which characterize the stability properties of the uniform states with respect to a dynamical system, which gives at least for large populations a good approximation of the behavior of the GA. In section 4.3 this system of difference equations is formulated and in section 4.4 the term "good" approximation is exactified.

4.3 The Difference Equations Describing the GA

In the following analysis we will no longer deal with the Markov process $\{\Phi_t\}$ but rather with a deterministic discrete time dynamical system, which is presumed to be a good approximation of the actual behavior of the stochastic process. Following Nix and Vose [133] the state of the system in period $t+1$ is defined as the conditional expected value of the state of the Markov process at time $t+1$ given the current state in period t. This gives a dynamical system of the form

$$\phi_{t+1} = \mathcal{G}(\phi_t), \tag{4.3}$$

where $\mathcal{G}$ is defined by

$$\mathcal{G}(\phi) = \mathbb{E}(\Phi_{t+1}|\Phi_t = \phi),$$

and where $\{\Phi_t\}$ is the Markov process considered in section 4.2. Since we have a state dependent fitness function now, the operator $\mathcal{G}$ differs from the one mentioned in chapter 3. The state space of the difference equation is the limit of S for $n \to \infty$, namely the whole r-dimensional simplex Δ^r. Thus we have to assume that the fitness function $f(\phi)$ is defined on the whole simplex and not only on some discrete subset $S \subset \Delta^r$. In the further analysis f is always assumed to be continuously differentiable.

In order to calculate $\mathcal{G}$, note first that the expected number of strings k in the population is $n/2$ times the expected number of offspring k from one arbitrary application of crossover and mutation. Denote by $m : \Omega \to \Omega$ the mutation operator and by $c : \Omega \times \Omega \to \Omega \times \Omega$ the crossover operator, and write $p_i(\phi)$ for the probability that the i-th individual is selected from a population in state ϕ. Then

$$\begin{aligned}
&\mathbb{E}(\text{number of offspring } k|\phi) \\
&= \sum\nolimits_{i,j\in\Omega} \mathbb{E}(\text{number of offspring } k|\text{parents are } i \text{ and } j)\, p_i(\phi)p_j(\phi) \\
&= \sum\nolimits_{i,j\in\Omega} \mathbb{E}(I_{\{k\}}(m(c_1(i,j))) + I_{\{k\}}(m(c_2(i,j))))\, p_i(\phi)p_j(\phi) \\
&= \sum\nolimits_{i,j\in\Omega} (\mathbb{P}(m(c_1(i,j)) = k) + \mathbb{P}(m(c_2(i,j)) = k))\, p_i(\phi)p_j(\phi)
\end{aligned}$$

$$= \sum_{i,j \in \Omega} 2m_{ij}(k) p_i(\phi) p_j(\phi),$$

where I is the indicator function, and

$$m_{ij}(k) = \frac{1}{2}\mathbb{P}(m(c_1(i,j)) = k) + \frac{1}{2}\mathbb{P}(m(c_2(i,j)) = k)$$

is the probability that an arbitrary offspring of the crossover between string $i \in \Omega$ and string $j \in \Omega$ after mutation is the string $k \in \Omega$.

The expected relative number of offspring k in a population P in state ϕ is therefore given by

$$\mathbb{E}([\Phi_{t+1}]_k | \Phi_t = \phi) = \mathcal{S}(\phi)^T M(k) \mathcal{S}(\phi),$$

where

$$\mathcal{S}(\phi) = \frac{\mathrm{diag}(f(\phi))\phi}{f(\phi)^T \phi} \tag{4.4}$$

is the selection operator incorporating proportional selection with a state-dependent fitness function. Using the notation of subsection 3.5.3 the above expression may be written as

$$\mathbb{E}([\Phi_{t+1}]_k | \Phi_t = \phi) = \mathcal{M}_k(\mathcal{S}(\phi)).$$

Therefore, we get finally

$$\mathcal{G}(\phi) = \mathcal{M}(\mathcal{S}(\phi)), \tag{4.5}$$

where $\mathcal{M}$ is given by (3.2) and $\mathcal{S}$ by (4.4).

The functional form of (4.4) is the form of the well known replicator dynamics[3], which is of great importance in the fields of biology and evolutionary game theory. The dynamic features of the discrete time replicator dynamics are quite complex and until now not fully understood. As the dynamics in equation (4.3) is basically a composition of the replicator dynamics and a non-linear mixing operator it is to be anticipated that the dynamics of the GA will not be easy to analyze.

In order to justify the approach to analyze the trajectory of (4.3) instead of the underlying Markov process I show analytically in section 4.4 and numerically in section 4.5 that the deviation is with high probability negligible if the population is sufficiently large.

4.4 Deviation from the Markov Process

In the last section the operator $\mathcal{G}(\phi)$, which describes the expected state of the successor of a population in state ϕ (note however that $\mathcal{G}^t(\phi)$ is not necessarily the expected state of ϕ_t given that $\Phi_0 = \phi$) was introduced. One

[3] We deal with the replicator dynamics in chapter 5

would expect that the trajectory $\{\mathcal{G}^t(\Phi_0)\}$ is with a high probability a good approximation for the original Markov process, provided that n is sufficiently large. In proposition 4.4.1 this intuition is exactified by showing that for $n \to \infty$ the maximal deviation on every finite time interval converges to zero in probability.

Proposition 4.4.1. *Let Φ_t be the trajectory of the Markov chain modelling an SGA and let $\mathcal{G}(\phi)$ be defined by (4.5). Then for $n \to \infty$, $\{\Phi_t\}$ converges in probability to $\{\mathcal{G}^t(\Phi_0)\}$, uniformly on every finite time interval $[0, T]$.*

Proof. Let $\|\cdot\|$ be an arbitrary norm on $\mathbb{R}^r$. We have to show that for every $T, \epsilon > 0$ and $0 < \gamma < 1$ there exists an N such that for all $n > N$,

$$\mathbb{P}\left(\max_{0 \le t \le T} \|\Phi_t - \mathcal{G}^t(\Phi_0)\| > \epsilon\right) < \gamma.$$

It is easy to see that the operator $\mathcal{G}$ is continuously differentiable on the r-dimensional simplex. As the simplex is compact, $\mathcal{G}$ is Lipschitz continuous on it with some Lipschitz constant $L > 0$. Define $\epsilon_t := \epsilon(L+1)^{t-T}$. Obviously, $0 < \epsilon_t \le \epsilon$ for all $t \le T$. We have

$$\begin{aligned}
&\mathbb{P}\left(\max_{0 \le t \le T} \|\Phi_t - \mathcal{G}^t(\Phi_0)\| \le \epsilon\right) \\
&\ge \mathbb{P}(\|\Phi_1 - \mathcal{G}(\Phi_0)\| < \epsilon, \ldots, \|\Phi_T - \mathcal{G}^T(\Phi_0)\| < \epsilon) \\
&\ge \mathbb{P}(\|\Phi_1 - \mathcal{G}(\Phi_0)\| < \epsilon_1, \ldots, \|\Phi_T - \mathcal{G}^T(\Phi_0)\| < \epsilon_T) \\
&\ge \prod_{t=1}^{T} \inf_{\phi \in \Delta^r : \|\mathcal{G}^{t-1}(\Phi_0) - \phi\| < \epsilon_{t-1}} \mathbb{P}(\|\Phi_t - \mathcal{G}^t(\Phi_0)\| < \epsilon_t | \Phi_{t-1} = \phi),
\end{aligned}$$

where the last inequality holds because $\{\Phi_t\}$ is a Markov chain, and all other inequalities are trivial.

Next, observe that, given Φ_{t-1}, $n\Phi_t$ is the sum of $n/2$ identically distributed independent random vectors describing the result of selection of two strings, their crossover and mutation. As each component of such a random vector is smaller or equal to 2, we have $\|V\|_\infty < 4$, where V is the variance-covariance matrix of one such random vector given Φ_{t-1}. Note that V depends on Φ_{t-1}, but that the above inequality holds for all states Φ_{t-1} in Δ^r. As the conditional expectation of Φ_t given Φ_{t-1} is $\mathcal{G}(\Phi_{t-1})$, we obtain from Chebyshev's inequality that for arbitrary $\tau > 0$ and $\phi \in \Delta^r$,

$$\mathbb{P}(\|\Phi_t - \mathcal{G}(\Phi_{t-1})\|_\infty > \tau | \Phi_{t-1} = \phi) < \frac{2}{n\tau^2}.$$

As all norms on finite-dimensional spaces are equivalent, this implies that for every $0 \le t \le T$ there exists an integer $N(t)$ such that

$$\mathbb{P}(\|\Phi_t - \mathcal{G}(\Phi_{t-1})\| < \epsilon_{t-1} | \Phi_{t-1} = \phi) > \sqrt[T]{1-\gamma}$$

for all $n > N(t)$ and all states ϕ in Δ^r.

Using these facts we find that for any $\phi \in \Delta^r$ with $||\mathcal{G}^{t-1}(\Phi_0) - \phi|| < \epsilon_{t-1}$,

$$\begin{aligned}
&\mathbb{P}(||\Phi_t - \mathcal{G}^t(\Phi_0)|| < \epsilon_t | \Phi_{t-1} = \phi) \\
&\geq \mathbb{P}(||\Phi_t - \mathcal{G}(\Phi_{t-1})|| + ||\mathcal{G}(\Phi_{t-1}) - \mathcal{G}^t(\Phi_0)|| < \epsilon_t | \Phi_{t-1} = \phi) \\
&\geq \mathbb{P}(||\Phi_t - \mathcal{G}(\Phi_{t-1})|| < \epsilon_t - L\epsilon_{t-1} | \Phi_{t-1} = \phi) \\
&= \mathbb{P}(||\Phi_t - \mathcal{G}(\Phi_{t-1})|| < \epsilon_{t-1} | \Phi_{t-1} = \phi) \\
&> \sqrt[T]{1-\gamma}
\end{aligned}$$

for all $n > N(t)$. Hence,

$$\inf_{\phi \in \Delta^r : ||\mathcal{G}^{t-1}(\Phi_0) - \phi|| < \epsilon_{t-1}} \mathbb{P}(||\Phi_t - \mathcal{G}^t(\Phi_0)|| < \epsilon_t | \Phi_{t-1} = \phi) \geq \sqrt[T]{1-\gamma}.$$

With $N := \max_{0 \leq t \leq T} N(t)$ we finally obtain

$$\mathbb{P}\left(\max_{0 \leq t \leq T} ||\Phi_t - \mathcal{G}^t(\Phi_0)|| \leq \epsilon\right) > 1 - \gamma$$

for all $n > N$, whence the proposition.

As the process $\mathcal{G}^t(\Phi_0)$ is deterministic the convergence in probability is equivalent to the weak convergence of the GA trajectory to the solution of (4.3) (see Billingsley [27], p.24f).

Obviously the operator $\mathcal{G}$ depends on the mutation probability μ. Sometimes it is convenient to emphasize this dependence by writing $\mathcal{G} = \mathcal{G}_\mu$. In our analysis of the stability of the uniform states we will — for mathematical reasons — not deal with the general expression for $\mathcal{G}_\mu(\phi)$, but will rather analyze the special case $\mathcal{G}_0(\phi)$. In most practical simulations the mutation probability will however be positive, and a positive μ seems to be also perfectly reasonable in the context of an economic learning model. On the other hand, the mutation probability usually is very small and in proposition 4.4.2 it is shown that the deviation of $\{\Phi_t\}$ from $\{\mathcal{G}_0^t(\Phi_0)\}$ is with a high probability small if μ is small and n is large.

Proposition 4.4.2. *Let $\{\Phi_t\}$ be the Markov process modelling an SGA with mutation probability $\mu > 0$ and population size n. Then, if $n \to \infty$ and $\mu \to 0$, $\{\Phi_t\}$ converges in probability to $\{\mathcal{G}_0^t(\Phi_0)\}$, uniformly on every finite time interval* $[0, T]$.

Proof. The proof is similar to that of proposition 4.4.1. Fix $T, \epsilon > 0$ and $0 < \gamma < 1$. Let $L > 1$ be a Lipschitz constant of $\mathcal{G}_0$ on Δ^r, and let $\delta = \frac{\epsilon(1-L)}{2(1-L^T)} > 0$ and $\alpha = \delta/2$.

We start by analyzing the sensitivity of $\mathcal{G}$ with respect to μ. As we know from chapter 3, the i-th component of this operator is given by

$$[\mathcal{G}_\mu(\phi)]_i = (\sigma_i \mathcal{S}(\phi))^T M_\mu (\sigma_i \mathcal{S}(\phi)),$$

where M_μ is an $r \times r$ matrix whose elements are polynomials in μ. This implies that the derivative of $\mathcal{G}$ with respect to μ is continuous on $\Delta^r \times [0,1]$. Hence, there is a $\tilde{\mu}$ such that

$$||\mathcal{G}_\mu(\phi) - \mathcal{G}_0(\phi)|| < \delta \;\; \forall \mu < \tilde{\mu}, \; \forall \phi \in \Delta^r.$$

Next, we define a sequence $\{\epsilon_t\}$ by the recursion

$$\begin{aligned} \epsilon_0 &= \epsilon/(4L^T), \\ \epsilon_{t+1} &= \alpha + \delta + L\epsilon_t, \quad t \geq 0. \end{aligned}$$

Simple calculations show that $\{\epsilon_t\}$ is an increasing sequence with $0 < \epsilon_t \leq \epsilon$. Similar to the proof of proposition 4.4.1, it can be seen that there is an integer $\tilde{N}$ such that

$$\mathbb{P}(||\Phi_t - \mathcal{G}_\mu(\Phi_{t-1})|| < \alpha | \Phi_{t-1} = \phi) > \sqrt[T]{1-\gamma}$$

for all $0 \leq t \leq T$, $\mu > 0$ and $\phi \in \Delta^r$, provided that $n > \tilde{N}$.

Putting these facts together, we obtain that for any $\phi \in \Delta^r$ with $||\mathcal{G}_0^{t-1}(\Phi_0) - \phi|| < \epsilon_{t-1}$,

$$\begin{aligned} &\mathbb{P}(||\Phi_t - \mathcal{G}_0^t(\Phi_0)|| < \epsilon_t | \Phi_{t-1} = \phi) \\ &\geq \mathbb{P}(||\Phi_t - \mathcal{G}_\mu(\Phi_{t-1})|| + ||\mathcal{G}_\mu(\Phi_{t-1}) - \mathcal{G}_0(\Phi_{t-1})|| \\ &\qquad + ||\mathcal{G}_0(\Phi_{t-1}) - \mathcal{G}_0^t(\Phi_0)|| < \epsilon_t | \Phi_{t-1} = \phi) \\ &\geq \mathbb{P}(||\Phi_t - \mathcal{G}_\mu(\Phi_{t-1})|| + \delta + L\epsilon_{t-1} < \epsilon_t | \Phi_{t-1} = \phi) \\ &= \mathbb{P}(||\Phi_t - \mathcal{G}_\mu(\Phi_{t-1})|| < \alpha | \Phi_{t-1} = \phi) \\ &> \sqrt[T]{1-\gamma} \end{aligned}$$

for all $n > \tilde{N}$ and $\mu < \tilde{\mu}$, which in turn yields that for all $n > \tilde{N}$ and $\mu < \tilde{\mu}$,

$$\mathbb{P}\left(\max_{0 \leq t \leq T} ||\Phi_t - \mathcal{G}_0^t(\Phi_0)|| > \epsilon \right) < \gamma,$$

whence the proposition.

As a corollary we obtain the weak convergence of $\{\Phi_t\}$ to $\{\mathcal{G}_0^t(\Phi_0)\}$ on every finite interval.

The last two propositions give us a strong argument in favor of a deterministic analysis. In principle it is possible to calculate with the help of the arguments given above a population size n and a mutation probability μ which guarantee that the process will let us say with probability 0.95 stay for a given time period in a given neighborhood of our deterministic systems. Of course the actual calculations needed for such a statement are very tiresome and in general not worth doing. The important point is that if we use a GA with large population and small mutation probability the trajectory of the GA has with high probability the same qualitative behavior as the dynamical system $\mathcal{G}^t(\Phi_0)$. Thus, stability results for the deterministic system tell us a lot about the "stability" of different states with respect to the Markov process defined by the GA.

4.5 A Numerical Example

In order to illustrate proposition 4.4.1, I present at this point a simulation where the trajectories of the real GA and the deterministic approximation can be compared directly. Consider a situation where the prisoner's dilemma defined by the payoff matrix as in equation (2.1) with $T = 9, R = 6, P = 3, S = 1$ is played within a population. In each time period, each member of the population plays against all other members, or, in other words, every agent plays against the population distribution. Agents play mixed strategies, $(m, 1-m)$ and each agent is represented by a binary string. A very simple coding of a mixed strategy is used where the number $m \in [0,1]$ is encoded in a binary string of length 6. The pure strategy "defect", which corresponds to $(0,1)$, is therefore represented by the string 0, where all bit values are zero. In this framework I will compare the results of a simulation done with a simple genetic algorithm incorporating proportional selection, one-point crossover and mutation with the trajectory of the system of difference equations (4.3). The population size n is 2000, the crossover probability is 1, and the mutation probability 0.001. The state space is 64-dimensional but only two projections of the trajectories are depicted. In figure 4.1 the evolution of the populations average strategy $\bar{m}_t$, given by

$$\bar{m}_t = \sum_{i=0}^{63} m(i)(\phi_t)_i,$$

is shown, where $m(i)$ is the real number encoded by the binary string i.

In order to calculate the expected trajectory of the GA according to (4.3), we first have to calculate the mixing matrix M for the case of one-point crossover. Recall that m_{ij} denotes the probability that an arbitrary offspring of the parents i and j is the string 0. Let us denote by $|i|$ the number of bits values 1 in string i, and by $|i,n|$ the number of bit values 1 in the first n bits of the string i. Assume first that no crossover occurs. In this case an arbitrary offspring from crossover of the parents i and j is with probability $\frac{1}{2}$ i and with probability $\frac{1}{2}$ the string j. In the first instance we need exactly $|i|$ mutations to construct the string 0, in the second we need $|j|$ mutations. The probability for the event that no crossover occurs and an arbitrarily chosen offspring equals 0 is therefore given by

$$\frac{1-\chi}{2}[\mu^{|i|}(1-\mu)^{l-|i|} + \mu^{|j|}(1-\mu)^{l-|j|}].$$

Now let us assume that crossover occurs and the crossover point is $p \in \{1, \ldots, l-1\}$. In this case we need either $|i,n|$ mutations in the first p bits and $|j| - |j,n|$ in the rest of the string, or $|j|$ plus $|i| - |i,n|$ mutations. The probability for this event is given by

$$\frac{\chi}{2(l-1)}[\,\mu^{|i,n|}(1-\mu)^{p-|i,n|}\mu^{|j|-|j,n|}(1-\mu)^{(l-p)-(|j|-|j,n|)}$$
$$+\mu^{|j,n|}(1-\mu)^{p-|j,n|}\mu^{|i|-|i,n|}(1-\mu)^{(l-p)-(|i|-|i,n|)}].$$

Summing over all possible events which yield an offspring 0, we get after reordering of terms

$$m_{ij} = \tfrac{(1-\mu)^l}{2}\left[(1-\chi)\left[\left(\tfrac{\mu}{1-\mu}\right)^{|i|} + \left(\tfrac{\mu}{1-\mu}\right)^{|j|}\right]\right.$$
$$\left. + \tfrac{\chi}{l-1}\sum_{p=1}^{l-1}\left[\left(\tfrac{\mu}{1-\mu}\right)^{|j|+|i,n|-|j,n|} + \left(\tfrac{\mu}{1-\mu}\right)^{|i|+|j,n|-|i,n|}\right]\right]. \tag{4.6}$$

Using this formula the trajectory of the system (4.3) can be plotted for our example and compared with the results of a simulation with a real GA. The code for this simulation was written in FORTRAN and executed on a DEC 3000 workstation. The trajectory of the real GA and of the approximation by the difference equation are hardly distinguishable. 200 generations were calculated and we see very clearly the evolution toward the unique equilibrium $((0,1),(0,1))$ (figure 4.1). Figure 4.2 shows how the fraction of agents playing the pure strategy "defect" increases. The discrepancy in this figure is a little higher, but this must be expected, as the summations to compute the average strategy levels out the stochastic errors which can be quite clearly seen here. Nevertheless, these simulations are a strong confirmation of the analytically obtained insight that the solution of the system (4.3) is well suited to describing the behavior of a genetic algorithm in a system with a state-dependent fitness function.

4.6 Stability of the Uniform States

In the last two sections it was demonstrated that the trajectory generated by the nonlinear difference equation (4.3) is a good approximation of the actual behavior of the GA, provided that the size of the population is sufficiently large. On the other hand, we have seen in section 4.1 that in the long run the state of the population will with high probability be eventually uniform, if the mutation probability is sufficiently small. Putting these two facts together, it is natural to analyze which of the uniform states attract nearby trajectories and which uniform states repel them [4]. Obviously only the uniform states which are at least locally asymptotically stable, are possible limit points of the process. We will analyze the case where $\mu = 0$, but since the right hand side of (4.3) depends continuously on μ, the stability properties derived for $\mu = 0$ also hold for small $\mu > 0$. Note first that for $\mu = 0$ every uniform state e_k, $k = 0, \ldots, r-1$ is a fixed point of (4.3). In proposition 4.6.1 the spectrum of the linearization of (4.3) at a fixed point $e_k \in \Delta^r$ is calculated[5].

[4] See Appendix A.2 for the key results concerning stability in nonlinear difference equations. A general introduction to this topic may be found in Kelley and Peterson [94].

[5] Vose and Wright [177] have derived a similar proposition for the special case of fitness functions which do not depend on the population state.

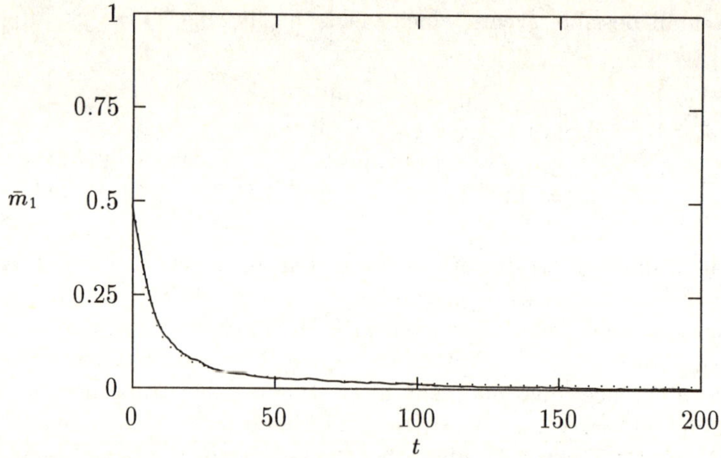

Fig. 4.1. The average population strategy in the iterated prisoner's dilemma according to the approximating deterministic dynamics and in a real simulation with a simple GA ($n = 2000,\ l = 6,\ \chi = 1,\ \mu = 0.001$)

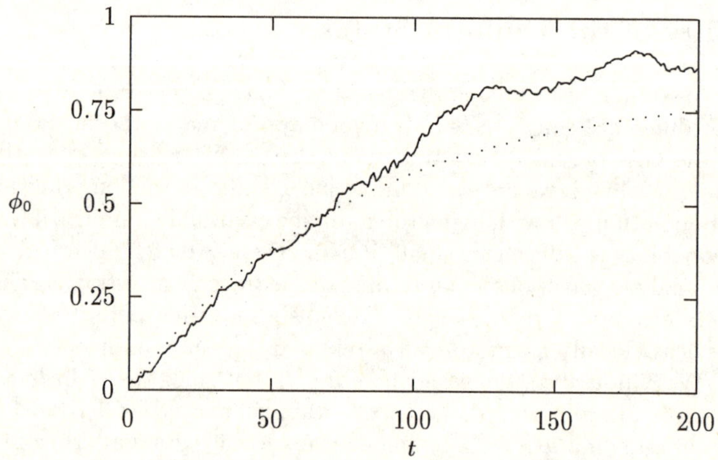

Fig. 4.2. The fraction of individuals playing the pure strategy "defect" according to the approximating deterministic dynamics and in a real simulation with a simple GA ($n = 2000,\ l = 6,\ \chi = 1,\ \mu = 0.001$)

Proposition 4.6.1. *For $\mu = 0$ the spectrum of the linearization of (4.3) at $e_k \in \Delta^r$ is given by*

$$\Lambda(e_k) = \{0\} \cup \left\{ 2m_{0,j\oplus k} \frac{f_j(e_k)}{f_k(e_k)}, \quad j \in \Omega,\ j \neq k \right\}. \tag{4.7}$$

The value 0 is the external eigenvalue[6] *and has no impact on the dynamics of the system within the simplex.*

Proof. We explicitly calculate the $r \times r$ matrix $\mathcal{D}_{\mathcal{M}\circ\mathcal{S}}(e_k) = \mathcal{D}_{\mathcal{M}}(e_k)\mathcal{D}_{\mathcal{S}}(e_k)$. As already mentioned in section 3.5.3, direct calculation yields $\mathcal{D}_{\mathcal{M}}(\phi) = 2\sum_{j=0}^{r-1} \sigma_j^{-1} M^* \sigma_j \phi_j$ with $M^* = [m_{i\oplus j,i}]_{i,j\in\Omega}$ (see Vose and Liepins [175]), and therefore $\mathcal{D}_{\mathcal{M}}(e_k) = 2\sigma_k^{-1} M^* \sigma_k$. Note further that M^* is an upper triangular matrix for $\mu = 0$. This can be seen by considering $m^*_{i,j} = m_{i\oplus j,i}$ for $i > j$. In this case there has to be a bit position p with $i(p) = 1$ and $j(p) = 0$, and therefore $(i \oplus j)(p) = 1 = i(p)$. This being so string 0 can never emerge from crossover between $i \oplus j$ and i, which implies $m^*_{i,j} = m_{i\oplus j,i} = 0\ \forall i,j \in \Omega$ with $i > j$. The linearization of the selection operator given by (4.4) reads:

$$\begin{aligned}
\mathcal{D}_{\mathcal{S}}(e_k) &= \tfrac{1}{(e_k^T f(e_k))^2}[(\mathrm{diag}(f(e_k)) + \mathrm{diag}(e_k)\mathcal{D}_f(e_k))(e_k^T f(e_k)) \\
&\quad -\mathrm{diag}(f(e_k))e_k(f^T(e_k) + e_k^T \mathcal{D}_f(e_k))] \\
&= \tfrac{1}{f_k{}^2(e_k)}[f_k(e_k)\mathrm{diag}(f(e_k)) + f_k(e_k)\mathrm{diag}(e_k)\mathcal{D}_f(e_k) \\
&\quad - \underbrace{\mathrm{diag}(f(e_k))e_k}_{=f_k(e_k)e_k} f_k^T(e_k) - \underbrace{\mathrm{diag}(f(e_k))e_k e_k^T}_{=f_k(e_k)\mathrm{diag}(e_k)} \mathcal{D}_f(e_k)] \\
&= \tfrac{1}{f_k(e_k)}[\mathrm{diag}(f(e_k)) - e_k f^T(e_k)]
\end{aligned}$$

Let D_k^* be given by

$$\begin{aligned}
D_k^* &= \sigma_k \mathcal{D}_{\mathcal{S}}(e_k)\sigma_k = \tfrac{1}{f_k(e_k)}[\underbrace{\sigma_k \mathrm{diag}(f(e_k))\sigma_k}_{=\mathrm{diag}(\sigma_k f(e_k))} - \underbrace{\sigma_k e_k}_{=e_0} f^T(e_k)\sigma_k] \\
&= \tfrac{1}{f_k(e_k)}[\mathrm{diag}(\sigma_k f(e_k)) - e_0 f^T(e_k)\sigma_k]
\end{aligned}$$

This means that D_k^* is the difference between a diagonal matrix and a matrix, which has positive values only in the first row. Therefore D_k^* is upper triangular and it is easy to see that the elements of the diagonal are given by

$$\begin{aligned}
d^*{}_{jj} &= 0 & j &= 0 \\
d^*{}_{jj} &= \tfrac{f_{j\oplus k}(e_k)}{f_k(e_k)} & j &= 1, \dots, r-1.
\end{aligned}$$

[6] Since the simplex is invariant under this dynamical system, all but one right eigenvectors of the linearization at an arbitrary point within the simplex lie in the hyperplane defined by $\sum_{k=0}^{r-1} x_k = 0$ and only one outside. The corresponding eigenvalue is called external eigenvalue and has no impact on the behavior of the dynamical system within the simplex.

Using ${\sigma_k}^{-1} = \sigma_k$ we get

$$\begin{aligned}\mathcal{D}_{\mathcal{M}\circ\mathcal{S}}(e_k) &= 2\sigma_k M^* \sigma_k \mathcal{D}_\mathcal{S}(e_k) = 2\sigma_k M^* \sigma_k \sigma_k D_k^* \sigma_k \\ &= 2\sigma_k M^* D_k^* \sigma_k.\end{aligned}$$

As a permutation of the basis does not change the eigenvalues of a matrix, the spectrum of $\mathcal{D}_{\mathcal{M}\circ\mathcal{S}}(e_k)$ coincides with the spectrum of $2M^* D_k^*$. As M^* and D_k^* are both upper triangular matrices, the eigenvalues of the product are given by the product of the diagonal elements of the two matrices, which yields

$$\Lambda(e_k) = \{0\} \cup \left\{ 2m_{0,i} \frac{f_{i\oplus k}(e_k)}{f_k(e_k)},\ i = 1, \dots r-1 \right\}.$$

Substituting $i \oplus k$ by j gives (4.7). To establish that 0 is the external eigenvalue, we show that the corresponding right eigenvector is given by e_k. Recalling that $\sigma_k e_k = e_0$ we get

$$\mathcal{D}_{\mathcal{M}\circ\mathcal{S}}(e_k) e_k = 2\sigma_k M^* D_k^* e_0 = 0,$$

since the first column of $M^* D_k^*$ contains only zeros. Further it is easy to realize that the vector e_k points out of the simplex Δ^r at the state e_k. Thus, 0 is the external eigenvalue and has no impact on the behavior of the considered dynamics.

Proposition 4.6.1 is a rather interesting result. Despite the huge state space and the highly nonlinear character of the difference equation system, we can explicitly compute the whole spectrum of the linearization. All eigenvalues are basically the product of some fitness ratio at e_k and a parameter determined by the kind of crossover used. We would like to point out that the eigenvalues depend only on the values of the fitness function but not on its derivatives. This highly surprising fact has a very positive effect. If we consider a GA with a large but finite population, it is quite obvious that only the function values of f on the finite set S may influence the behavior of the GA. Thus, all fitness functions whose values match on this set yield the same behavior. As our deterministic dynamical system is used as an approximation of the behavior of the GA, it would be very disturbing if function values at points which are no element of S or the derivatives of f had an influence in our analytical analysis. However, according to proposition 4.6.1 the spectrum of the process depends only on function values at uniform states. As uniform states are an element of S for any $n > 0$, the qualitative behavior of the dynamical system is indeed never influenced by properties of f which do not affect the GA.

As it is well known that a fixed point of a system of difference equations is locally asymptotically stable if all eigenvalues of the linearized system lie in the interior of the unit circle in the complex plane[7], proposition 4.6.1 is

[7] See e.g Kelley and Peterson [94].

very useful in determining the local stability properties of the uniform states. It enables us to derive stability results for different crossover operators by just calculating the term $m_{0,j\oplus k}$. Here I will do so for two of the most widely used crossover operators, namely, one-point crossover and uniform crossover.

To formulate the stability properties for one point crossover, let us denote by $d(j) = \max\{|p-q| \; |j(p) = j(q) = 1\}$ the longest distance between two bits with value 1 in string j.

Proposition 4.6.2. *A uniform state e_k is locally asymptotically stable for the expected dynamics of a GA (4.3) with $\mu = 0$ and one-point crossover with probability $\chi \in (0,1]$ if*

$$\frac{d(j \oplus k)}{l-1} > \frac{1}{\chi}\left(1 - \frac{f_k(e_k)}{f_j(e_k)}\right)$$

for all $j \in \Omega, j \neq k$. If there is one $j \neq k$ such that the inequality holds the other way round, e_k is unstable.

Proof. The state e_k is stable with respect to (4.3) if

$$2m_{0,j\oplus k}\frac{f_j(e_k)}{f_k(e_k)} < 1$$

for all $j \in \Omega,\ j \neq k$. Recall that $m_{0,j\oplus k}$ is the probability that an arbitrary chosen offspring of the parents 0 and $j \oplus k$ is 0. One of the two offspring is 0 if no crossover takes place, or if the crossover point is left of the leftmost 1 in $j \oplus k$ or right of the rightmost 1 in $j \oplus k$. This yields

$$m_{0,j\oplus k} = \frac{1}{2}\left(1 - \frac{\chi d(j \oplus k)}{l-1}\right).$$

Inserting this expression into the inequality above establishes the first part of proposition 4.6.2. The instability result follows directly from the fact that a fixed point of a dynamical system is unstable whenever one eigenvalue has a modulus larger than one.

Note that $d(j \oplus k)$ is the length between the two outmost bits where j and k differ in value. Proposition 4.6.2 therefore claims that a state consisting almost only of strings k will converge to the uniform state e_k if the strings receiving a higher payoff in the current state differ from k in bits positioned far apart. This can be explained by the fact that if few such strings j appear in the population they grow in number by selection, but there is still an overwhelming probability that they are paired for crossover with a string k (as long as the ratio of fitness values of j and k is not too large) and are destroyed by crossover with a high probability, as any crossover point position between the two outmost differing bits destroys j. Accordingly, an important implication of proposition 4.6.2 is that the encoding of a problem may determine the stability or instability of certain uniform states.

As to interpret this result with respect to the economic interpretation of a GA given in section 3.6, we might conclude that, due to communication effects, homogeneous states may be stable even if it is no economic equilibrium. If in such a population one individual changes his strategy to some other with a higher payoff in the current environment it depends on "how different" this strategy is from the prevalent population strategy whether the individual continues to use this strategy. If the difference is large then with high probability information exchange with some other individual in the population leads to a change of strategy. On the other hand, if the new strategy differs only in a few points, the chances that the new strategy is used in the future is large. Thus, information exchange between individuals has a stabilizing effect for almost uniform states where all strategies which pay better than the one mainly used differ a lot from this strategy. However, the exact meaning of "differing a lot" is not completely clear from an economic point of view as this depends on the coding mechanism and on the type of crossover operator being used. As already pointed out it is in general hard to give a meaningful economic interpretation of both of these technical details.

To demonstrate the dependence of the last result on the type of crossover operator, the next proposition is formulated under the assumption that uniform crossover is used in the GA, where at any position in the crossover mask 1 and 0 appear with the same probability 0.5.

Proposition 4.6.3. *A uniform state e_k is locally asymptotically stable for the expected dynamics of a GA (4.3) with $\mu = 0$ and uniform crossover with probability $\chi \in (0,1]$ if*

$$\frac{f_k(e_k)}{f_j(e_k)} > (1-\chi) + \chi \left(\frac{1}{2}\right)^{|j \oplus k|-1}$$

holds for all $j \in \Omega, j \neq k$. If there is one $j \neq k$ such that the inequality holds the other way round, e_k is unstable.

Proof. The proof is analogous to the proof of proposition 4.6.2, but of course the expression for $m_{0,j\oplus k}$ changes, for it is now assumed that uniform crossover is used. One of the two offspring of the parents 0 and $j \oplus k$ is the string 0, whenever no crossover occurs, or when the crossover mask has the same value at all bits, where the bit value of the string $j \oplus k$ is 1. The probability, that one offspring is 0, when uniform crossover is applied to the pair 0 and $j \oplus k$, is therefore given by $(1/2)^{|j\oplus k|-1}$. This yields

$$m_{0,j\oplus k} = \frac{1}{2}(1-\chi) + \chi \left(\frac{1}{2}\right)^{|j\oplus k|}.$$

Again we see that a uniform state e_k may be attractive, even if k is not the best reply to e_k, if all better replies are sufficiently far away. This time however the topology is different, and far away does not mean that two strings differ in bits far apart, but rather that they differ in many different bit positions. This implies that if uniform crossover is used, the stability properties of a uniform state can not be altered by just reordering the bits in the string, like it is the case with one-point crossover. With uniform crossover the coding has to be changed as such, if the stability of uniform states should be changed. It can be concluded that for every kind of crossover there is some kind of distance function, for which a similar result like propositions 4.6.2 and 4.6.3 holds.

Realizing this fact it is of course interesting to characterize the uniform states, whose stability is independent of the actual distance function, and which are therefore locally asymptotically stable (l.a.s.) for any kind of crossover used. Any kind of crossover here means any stochastic operator c which has the properties stated in chapter 3. Note that the operator which leaves both strings unchanged is formally a crossover operator and this implies that a state which is locally asymptotically stable for any kind of crossover must also be l.a.s. for a GA without crossover. We have realized already that non-equilibrium states may be stabilized by information exchange effects, but it is easy to realize that such a state can not be stable without this stabilizing effect. Proposition 4.6.4 shows that every non-equilibrium state is unstable for some kind of crossover operator but that, on the other hand, all strict economic equilibria are stable regardless of the crossover operator used. Remember that strict economic equilibria were introduced in definition 4.1.1.

Proposition 4.6.4. *Let e_k be a uniform strict equilibrium state of an SDF system defined by $f(\cdot)$, then e_k is locally asymptotically stable with respect to (4.3) for $\mu = 0$ and any kind of crossover operator used. On the other hand no uniform state which is no equilibrium is stable for all kinds of crossover operators.*

Proof. As e_k is a strict equilibrium, $f_k(e_k) > f_j(e_k)$ has to hold for any $j \neq k$. On the other hand, $m_{0,j} \leq \frac{1}{2}\ \forall j \neq 0$ holds for any crossover operator, since the crossover between the string 0 and any string not equal to 0 can never create two offspring 0. Together with (4.7) this yields that all eigenvalues of the linearization of (4.3) at e_k are strictly smaller than 1. If, on the other hand, e_k is no equilibrium state, there has to be a string $j \in \Omega$ with $f_j(e_k) > f_k(e_k)$. There is always some kind of crossover for which one of the offspring of 0 and $j \oplus k$ will be 0 with probability 1 (e.g. uniform crossover with a crossover mask consisting only of bits with value 0), which implies $m_{0,j\oplus k} = \frac{1}{2}$. According to proposition 4.6.1, the state e_k is not stable with respect to (4.3) if this kind of crossover is used, and therefore it can not be stable for all kind of crossover operators.

At this point it is interesting to compare this proposition with stability results concerning the replicator dynamics . It is well known that a uniform strict equilibrium is always l.a.s. with respect to the discrete time replicator dynamics whereas the Nash equilibrium property is necessary for the stability of a state (see e.g. Van Damme [169]). So, looking at proposition 4.6.4, we realize that the discrete time replicator dynamics – which exactly corresponds to the expected trajectory of a GA where no crossover and mutation operators but only proportional selection are applied – is the 'least stable' of all different GA dynamics stemming from the use of different crossover operators. This makes clear that any crossover operator facilitates the dynamic stability of uniform states.

Discussing proposition 4.6.2 we realized that the stability properties of a uniform state in general depend on both the type of crossover operator and the coding mechanism. Now, we have seen in proposition 4.6.4 that, by varying the type of the applied crossover operator, we can make any non-equilibrium state unstable. Proposition 4.6.5 shows that the same can be done by varying the coding mechanism. The proposition is formulated for one-point crossover, but similar results hold for any crossover operator.

Proposition 4.6.5. *Let e_k be a strict equilibrium state of an SDF system. Then e_k is locally asymptotically stable with respect to (4.3) for $\mu = 0$ and one-point crossover, regardless of the actual coding mechanism employed. Conversely, if e_k is not an equilibrium state, then there exists a coding mechanism such that e_k is unstable with respect to (4.3) for $\mu = 0$ and one-point crossover.*

Proof. If e_k is a strict equilibrium, $f_k(e_k) > f_j(e_k)$ has to hold for any $j \neq k$. Thus, the right hand side of the inequality in proposition 4.6.2 is negative which implies that the inequality is satisfied for any coding mechanism and e_k is always locally asymptotically stable. Conversely, if e_k is not an equilibrium state, there has to be a string $j \in \Omega$ with $f_j(e_k) > f_k(e_k)$. In fact, there is always a binary coding such that the Hamming distance between j and k is one. In this case, $d(j \oplus k) = 0$, which implies by proposition 4.6.2 that the state e_k is not stable with respect to (4.3) if this kind of coding is used with one-point crossover.

The results of propositions 4.6.4 and 4.6.5 are somewhat comforting compared with propositions 4.6.2 and 4.6.3. Whereas the latter imply that also states which do not correspond to an economic equilibrium may attract the trajectory of the GA, the former guarantee that this effect will vanish if we either change the crossover operator or the coding mechanism in an appropriate way. Thus, the stability of strict economic equilibria is in some sense more robust than the stability of non-equilibrium states.

At this point I would like to shortly discuss how the results derived so far can be applied if there is no one-to-one correspondence between the phenotype and the genotype of a string, which means that the binary representation of a strategy is not unique. Recently Novkovic and Sverko [134] have used GAs with 'genetic waste' for market simulations. In such a GA model not all bit positions in the string influence the strategy of the agent but some bits –

the genetic waste – are irrelevant for the determination of the individuals action. Since the binary string encoding a certain action is never unique there can be no strict uniform equilibrium in such a setup and it is easy to see that the stability criterions for one-point and uniform crossover can never be satisfied it there are at least two non-adjacent genetic waste bits. So, convergence on a genotype level is hard to establish in such a model. However, if we look at the problem on a phenotype level we realize that by considering the dynamics on a restricted state space, where only the bits relevant for the determination of the strategy are taken into account, the stability criteria can be adjusted such that they work for this restricted bijective representation. In particular, this implies that a state where all individuals use a strictly optimal strategy is stable on a phenotype level even if genetic waste prevents the binary representation to become uniform. Crucial for this result is that, in cases where non-uniqueness in the representation stems from genetic waste, any crossover between two different binary strings which represent the same strategy yields two binary strings representing this strategy as offspring. If this does not hold, non-uniqueness of representation may prevent the convergence of the behavior induced by a GA towards a state where all agents use the same strictly optimal strategy. Again, we see how important the type of binary representation chosen can be.

Before we proceed with two population models let me shortly summarize the main findings of this analytical examination of GAs in SDF systems. We have seen that in the long run the state of the population is close to uniform most of the time. Further it has been demonstrated not only by an analytical argument, but also by a simulation, that the behavior of a genetic algorithm with a large population can be satisfactorily approximated by a system of r non-linear difference equations in the r-dimensional simplex. Conditions characterizing uniform states which are stable with respect to these dynamics were derived for the cases of one-point and uniform crossover and the generalization to other types of crossover operators were outlined. In particular we have seen that every uniform state corresponding to a strict equilibrium of the economic system, defining the fitness of the strings, is locally asymptotically stable with respect to difference equations, regardless of the crossover operator and coding mechanism being used. In the next two chapters I will present simulations done for game theoretic and economic models where the obtained theoretical results will not only help us to shape the algorithm in a useful way, but also to interpret and understand the results of the simulations. Chapter 7 will show how in a general setting the theoretical results obtained in this chapter may be used to facilitate the learning of economic equilibria in simulations.

4.7 Two-Population Models

In this section I will transfer the analysis carried out up to now to a slightly different setup. Let us now assume that the payoff of an individual in the population is not only influenced by the actions of the other individuals in his own population but also by the state of some other populations. The interaction with the agents in the other populations is limited to the effects on the fitness. No imitation or communication with other populations is possible. In biological applications such a situation might occur when we consider the interaction between different species or between males and females (see for example Cressman [40]). The fitness of an individual depends on the actions taken by all individuals, but selection and crossover takes place only within a species. Also in economic systems it may be very useful to consider two-population models. For example, in markets there are buyers and sellers whose strategies influence each others payoff. However, a buyer is only interested in buyers' strategies and therefore will only communicate with buyers. Of course he may also be interested in sellers' strategies in order to anticipate future prices, but in our model he does not have the necessary information and computational capacity to do so (see section 6.3). Other models where more than one populations are required are overlapping generations models (see section 6.2) in which only the individuals with the same age are in the same population but the payoff of an individual depends also on the actions of agents with different age. Multi-population models also occur if individual learning is described and every agent is represented by a whole population of strings (see section 3.6.2). For reasons of simplicity only models with two populations are considered here. Of course the length of the strings, the size of the population and also other GA parameters may differ between the two populations. In what follows a superscript always indicates which population a parameter or operator belongs to.

Let us first shortly examine how the results derived for the Markov chain model can be transferred to a two-population framework. A state of the Markov chain is given by the pair (ϕ_1, ϕ_2) where ϕ_1 is the state of population one and ϕ_2 is the state of population two. It is easy to see that this chain is irreducible as long as the mutation probability is positive. So, let us look again at the limit of the unique stationary distribution of the chain for decreasing μ. Considering the proof of proposition 4.2.1 we see that the only property of the fitness which was used there is that all fitness values are positive. So, under the assumption that both fitness functions in the two-population model are positive we may apply exactly the same arguments separately to both populations in order to show that for decreasing μ the weight of all non-uniform profiles in the stationary distribution goes to zero whereas the weight of all uniform profiles stays strictly positive (the exact proof is omitted here because it is basically the same as the one of proposition 4.2.1).

Proposition 4.7.1. *Let π be the stationary distribution of the Markov chain describing a GA in a two-population model, then*

(i) $\lim_{\mu\to 0} \pi_{(\phi^1,\phi^2)} = 0 \quad \forall(\phi^1,\phi^2) \neq (e_j,e_k) \ j,k = 0,\dots,r-1.$

(ii) $\lim_{\mu\to 0} \pi_{(\phi^1,\phi^2)} > 0 \quad \forall(\phi^1,\phi^2) = (e_j,e_k) \ j,k = 0,\dots,r-1.$

Obviously, similar results also holds for any multi-population model.

Having established this, let us now turn to the expected dynamics of the system. Let ϕ_t^1 and ϕ_t^2 denote the corresponding deterministic dynamical system. The fitness functions are given by $f^i : \Delta^{r^1} \times \Delta^{r^2} \mapsto \mathbb{R}^{r^i}$ $i = 1,2$, where again $r^i = 2^{l^i}$. The fitness depends on the state of the other population as well which implies that the result of selection depends as well on the state of the own as the other population, i.e. $\mathcal{S}^i : \Delta^{r^1} \times \Delta^{r^2} \mapsto \Delta^{r^i}$ $i = 1,2$. The mixing operator on the other hand depends only on the state of the own mating pool. Thus, we may write the expected evolution of both populations as the following dynamical system

$$\begin{aligned}(\phi^1,\phi^2)_{t+1} &= \mathcal{G}(\phi_t^1,\phi_t^2) \\ &= (\mathcal{M}^1(\mathcal{S}^1(\phi_t^1,\phi_t^2)), \mathcal{M}^2(\mathcal{S}^2(\phi_t^1,\phi_t^2))).\end{aligned} \tag{4.8}$$

The selection operator is now given by

$$\mathcal{S}^i(\phi^1,\phi^2) = \frac{\operatorname{diag}(f^i(\phi^1,\phi^2))\phi^i}{f^i(\phi^1,\phi^2)^T\phi^i} \quad i = 1,2 \tag{4.9}$$

and the mixing operators are again given by

$$\mathcal{M}^i(\phi) = \left((\sigma_0^i\phi)^T M^i(\sigma_0^i\phi), \dots, (\sigma_{r^i-1}^i\phi)^T M^i(\sigma_{r^i-1}^i\phi)\right)^T, \tag{4.10}$$

where the elements $m_{j,k}^i$ of M^i give the probability that in population i an arbitrary offspring of strings j and k equals 0 and the σ_j^i denote basis transformations in $\mathbb{R}^{r^i}$. Note that $\mathcal{S}^1(e_k,\phi^2) = e_k$ and $\mathcal{S}^2(\phi^1,e_k) = e_k$ holds for all $\phi^1 \in \Delta^{r^1}$, $\phi^2 \in \Delta^{r^2}$ which implies that all combinations of uniform states (e_j,e_k) $0 \le j \le r^1-1$, $0 \le k \le r^2-1$ are fixed points of the dynamical system (4.8). Although the structure of the operators is a little more complicated now than that in one population models, a similar analysis can be carried out again. First we derive the pendant to proposition 4.6.1:

Proposition 4.7.2. *For $\mu = 0$ the spectrum of the linearization of (4.8) at $(e_j,e_k) \in \Delta^{r^1} \times \Delta^{r^2}$ is given by*

$$\begin{aligned}\Lambda(e_j,e_k) = \{0\} &\cup \left\{2m_{0,i\oplus j}^1 \tfrac{f_i^1(e_j,e_k)}{f_j^1(e_j,e_k)},\ i \in \Omega^1, i \neq j\right\} \\ &\cup \left\{2m_{0,i\oplus k}^2 \tfrac{f_i^2(e_j,e_k)}{f_k^2(e_j,e_k)},\ i \in \Omega^2, i \neq k\right\},\end{aligned} \tag{4.11}$$

where 0 is a double external eigenvalue.

Proof. The proof is quite similar to that of proposition 4.6.1. Denote by $\mathcal{D}_i$ $i = 1,2$ the derivative with respect to ϕ^i. From the proof of proposition 4.6.1 we get

$$\mathcal{D}_{1\mathcal{M}^1 \circ \mathcal{S}^1}(e_j, e_k) = 2\sigma_j^1 M^{1*} D_{jk}^{1*} \sigma_j^1,$$

where $M^{i*} = [m^i_{j \oplus k, j}]_{j,k \in \Omega^i}$ and

$$D_{jk}^{1*} = \frac{1}{f_j^1(e_j, e_k)}[\text{diag}(\sigma_j^1 f^1(e_j, e_k)) - e_0 f^1(e_j, e_k)^T \sigma_j^1].$$

Analogous to the proof of proposition 4.6.1 we realize that $M^{1*} D_{jk}^{1*}$ is an upper triangular matrix with diagonal elements

$$\{0\} \cup \left\{ 2m^1_{0, i \oplus j} \frac{f_i^1(e_j, e_k)}{f_j^1(e_j, e_k)},\ i \in \Omega^1, i \neq j \right\}.$$

In the same way we derive that

$$\mathcal{D}_{2\mathcal{M}^2 \circ \mathcal{S}^2}(e_j, e_k) = 2\sigma_k^2 M^{2*} D_{jk}^{2*} \sigma_k^2$$

and that $M^{2*} D_{jk}^{2*}$ is an upper triangular matrix with diagonal elements

$$\{0\} \cup \left\{ 2m^2_{0, i \oplus k} \frac{f_i^2(e_j, e_k)}{f_k^2(e_j, e_k)},\ i \in \Omega^2, i \neq k \right\}.$$

For $\mathcal{D}_{2\mathcal{S}^1}(e_j, e_k)$ we get

$$\begin{aligned}
&\mathcal{D}_{2\mathcal{S}^1}(e_j, e_k) \\
&= \frac{1}{f_j^1(e_j, e_k)^2}[\text{diag}(e_j)\mathcal{D}_{2f^1}(e_j, e_k) f_j^1(e_j, e_k) \\
&\qquad - \underbrace{\text{diag}(f^1(e_j, e_k)) e_j e_j^T}_{f_j^1(e_j, e_k)\text{diag}(e_j)} \mathcal{D}_{2f^1}(e_j, e_k) \\
&= 0,
\end{aligned}$$

and due to symmetry we also have $\mathcal{D}_{1\mathcal{S}^2}(e_j, e_k) = 0$. Together we have

$$\begin{aligned}
&\mathcal{D}_{\mathcal{G}}(e_j, e_k) \\
&= \begin{pmatrix} 2\sigma_j^1 M^{1*} D_{jk}^{1*} \sigma_j^1 & 0 \\ 0 & 2\sigma_k^2 M^{2*} D_{jk}^{2*} \sigma_k^2 \end{pmatrix} \\
&= 2 \begin{pmatrix} \sigma_j^1 & 0 \\ 0 & \sigma_k^2 \end{pmatrix} \begin{pmatrix} M^{1*} D_{jk}^{1*} & 0 \\ 0 & M^{2*} D_{jk}^{2*} \end{pmatrix} \begin{pmatrix} \sigma_j^1 & 0 \\ 0 & \sigma_k^2 \end{pmatrix}.
\end{aligned}$$

It is easy to see that the first and third matrix incorporate only a permutation of the basis of the $\mathbb{R}^{r^1+r^2}$ and the matrix in-between is an upper triangular matrix. Finally, it can be seen by analogous arguments to the ones given in the proof of proposition 4.6.1, that $\{(e_j, 0)', (0, e_k)'\}$ is a basis of the eigenspace of the double eigenvalue 0. Both of these vectors point out of the product of the simplices $\Delta^{r^1} \times \Delta^{r^2}$, which implies that 0 is the external eigenvalue and has no impact on the behavior of the dynamical system within $\Delta^{r^1} \times \Delta^{r^2}$. Hence the proposition.

Again, as in one population models, the eigenvalues of the linearization of the dynamics of the GA near a state where both populations are uniform are a product of two terms: the first one gives the probability that crossover between some string i and the prevalent string disrupts string i and the second one gives the ratio of the fitness values of these two strings in the current environment. The first of the two values depends only on the state of the own population and the second on the state of both populations. So again the effect of selection and mixing on the eigenvalues can be separated very well. With this result we get the following stability criterion for products of uniform states.

Proposition 4.7.3. *A state $(e_j, e_k) \in \Delta^{r^1} \times \Delta^{r^2}$ is locally asymptotically stable with respect to the expected dynamics of a two-population GA (4.8) with $\mu = 0$ and one-point crossover with probability $\chi \in (0,1]$ if*

$$\frac{d(i \oplus j)}{l^1 - 1} > \frac{1}{\chi}\left(1 - \frac{f_j^1(e_j, e_k)}{f_i^1(e_j, e_k)}\right) \qquad \forall i \in \Omega^1 \; i \neq j$$

$$\text{and} \quad \frac{d(i \oplus k)}{l^2 - 1} > \frac{1}{\chi}\left(1 - \frac{f_k^2(e_j, e_k)}{f_i^2(e_j, e_k)}\right) \qquad \forall i \in \Omega^2 \; i \neq k.$$

If there is some string i either in Ω^1 or in Ω^2 such that the corresponding inequality holds the other way round (e_j, e_k) is unstable.

Proof. Using the arguments given in the proof of proposition 4.6.2 the results follow directly from proposition 4.7.2.

Of course, the proposition 4.6.3 may also be easily formulated for two-population setups but this exercise is left to the reader. In order to analyze the stability properties of economic equilibria in this setup, a quite obvious extension of the definition of an economic equilibrium state to two-population models is used.

Definition 4.7.1. *A state $(\phi^1, \phi^2) \in S^1 \times S^2$ corresponds to an equilibrium of the economic system defined by $f^1(\cdot,\cdot)$ and $f^2(\cdot,\cdot)$ if and only if*

$$f_j^1(\phi^1, \phi^2) \geq f_i^1(\phi^1, \phi^2) \quad \forall i, j \in \Omega^1 \text{ with } \phi_j^1 > 0$$

$$f_k^2(\phi^1, \phi^2) \geq f_i^2(\phi^1, \phi^2) \quad \forall i, k \in \Omega^2 \text{ with } \phi_k^2 > 0.$$

If the first inequality holds strictly for all $i \neq j$ and the second for all $i \neq k$ the equilibrium is called strict.

Note again that a strict economic equilibrium has to be a state where both populations are uniform.

Using this definition we realize that also in two-population models states which are no economic equilibrium may be locally stable with respect to the expected dynamics of the GA. The explanation of this fact is basically the same as in one-population models. If in all populations all strings which have a higher fitness than the prevalent strings in their population differ from these strings in such a way that they are with high probability disrupted by crossover the corresponding non-equilibrium state may be stable. On the other hand we have

Proposition 4.7.4. *Let (e_j, e_k) be a strict equilibrium of the economic system defined by $f^1(\cdot,\cdot)$ and $f^2(\cdot,\cdot)$, then (e_j, e_k) is locally asymptotically stable with respect to the expected dynamics of the GA (4.8) with $\mu = 0$ regardless of the type of crossover and coding mechanism used. On the other hand, for any non-equilibrium state there is a crossover operator and a coding mechanism such that this state is unstable.*

Proof. The proof is analogous to that of proposition 4.6.4.

This section shows that the whole analysis carried out for one-population models applies with only minor differences also to two-population models. If we consider models where the fitness of a string depends on more than two populations the structure of the dynamics remains exactly the same. For all such models the linearization of the dynamics near states where all populations are uniform can be calculated explicitly by using the technique applied in this chapter and the spectrum will basically always look the same. The analytical results derived here will be used in the analysis of adaptive learning behavior in double auctions in section 6.3.

5. Genetic Learning in Evolutionary Games

5.1 Equilibria and Evolutionary Stability

Whereas the previous chapter was dedicated to the analytical examination of the dynamic properties of GAs in SDF systems, the following two chapters will deal with different specific applications. The purpose is to demonstrate how the combination of numerical experiments and the application of mathematical theory may lead to a sound understanding of the process in certain models. In this chapter we deal with GA learning in evolutionary games[1]. The theory of evolutionary games was first developed for biological models, but has attracted more and more attention of economists in the last few years. It deals with situations where individuals get some payoff from their interaction with other members of the same population. In general, it is assumed that all individuals have the same set of strategies at their disposal, and that the payoff they receive depends only on the own strategy and the opponent's strategy. For two-player games the payoffs can be written down in a payoff matrix and a game given by the set of strategies and the payoff matrix is called a *normal form game.* A game is an *evolutionary game* if the payoff of an individual is independent from the fact whether he is the first or the second player in the game. Thus, for every evolutionary game the payoff matrix of the second player is the transpose of the payoff matrix of the first player. In some cases it is assumed that each individual meets only one opponent in each period, where the matching is done randomly. Another possible setup is that every individual plays all other individuals in each period. In what follows it is assumed that the second setup holds, which implies that each individual plays against a virtual player who plays a mixed strategy corresponding to the population distribution.

Now, let A be the payoff matrix of a two person normal form game. The set of pure strategies is given by I, with $|I| = N$, and a mixed strategy m may be written as an N-tuple, with $m_s \geq 0 \; \forall s \in I$ and $\sum_{s \in I} m_s = 1$, where m_s is the probability that the pure strategy s is played. Let Δ^N denote the set of all mixed strategies. If we think of equilibrium concepts for this kind of

[1] Only a very short introduction to the basics of evolutionary game theory is given here. The interested reader is referred to Fudenberg and Tirole [63], Van Damme [169] or Weibull [180] for a more comprehensive coverage of this topic.

games, there is of course the concept of Nash-equilibria. We have to realize that, since the game is played within one population, we may only consider pairs of mixed strategies $(m^1, m^2) \in \Delta^N \times \Delta^N$ with $m^1 = m^2$. Therefore, we only deal with symmetric Nash-equilibria which are defined as follows:

Definition 5.1.1. *A strategy pair (m, m) is called a symmetric Nash equilibrium of the symmetric normal form game (A, A^T), iff*

$$m^T Am \geq \tilde{m}^T Am \; \forall \tilde{m} \in \Delta^N. \tag{5.1}$$

According to the famous theorem by Nash [131], every normal form game has at least one Nash equilibrium, and the same arguments as used in the proof of this theorem show that in the case of a symmetric game there has to be also a symmetric equilibrium (Van Damme [169]).

Next, assume that the population playing the game has already found a Nash equilibrium state. Assume further that there appear some mutants in the population, which slightly alter the population distribution. If this small deviation from the equilibrium path leads to a complete disruption of the equilibrium, the Nash equilibrium is highly unstable, and it is not very likely that the population will stay in this equilibrium state very long. On the other hand, if the population distribution returns to the equilibrium state after every small disruption, the equilibrium is stable with respect to small mutations, or — as it is called in the literature — evolutionary stable.

Definition 5.1.2. *A symmetric Nash equilibrium (m, m) is called an evolutionary stable strategy(ESS) iff there exists an $\epsilon^* > 0$ for every $\tilde{m} \in \Delta^N$, $\tilde{m} \neq m$, with*

$$m^T A((1-\epsilon)m + \epsilon\tilde{m}) > \tilde{m}^T A((1-\epsilon)m + \epsilon\tilde{m}) \;\; \forall \; \epsilon \leq \epsilon^*.$$

If the above condition holds, every invasion of mutants is most probably repulsed, because the majority of equilibrium strategy players has a higher fitness than the invaders, if the number of mutants is sufficiently small.

5.2 Learning in Evolutionary Games

The situation where an evolutionary game is played within a population has attracted a lot of attention from many researchers working in the field of learning dynamics. The question that is addressed most often is, which kind of learning dynamics leads to convergence towards an equilibrium and which kind of equilibria are learned. Various approaches to model the learning process have been adopted by different researchers, and I will shortly review the most important amongst them[2].

[2] The interested reader should have a look at the books of Fudenberg and Levine [62], Hofbauer and Sigmund [86], Samuelson [147] or Weibull [180].

Maybe the first learning concept used in game theory was that of fictitious play. The idea of this process is that every player builds up a belief about the mixed strategy of his opponent. The updating of the expected opponent's strategy is very simple. After having observed the action of the other player in period t, the a posteriori expectation of the opponent's mixed strategy is the weighted sum of the a priori expectation and the observed action with weights $\frac{t-1}{t}$ and $\frac{1}{t}$. Every player plays the strategy which is a best reply to this expected mixed strategy of the opponent. Note that the concept of fictitious play is a best response dynamics, and uses the assumption that all players know the payoff function of the game. Fictitious play has been analyzed in great detail. Robinson [142] showed that the beliefs generated by fictitious play converge in any finite two-person zero-sum game. For any non zero sum game where players choose between only two pure strategies, Miyasawa [123] has proved the convergence of fictitious play. Shapley [157] however, demonstrated by a counterexample that this result would not be necessarily true in games with more than two pure strategies. Krishna [104] has shown that fictitious play will find an equilibrium in any games with strategic complementarities, if some plausible assumptions about the payoff function are made.

A generalization of fictitious play, called Bayesian learning, has also attracted some attention. In this model it is assumed that the players hold not only an expectation about the strategy of their adversary, but that they also have a personal probability distribution about the possible mixed strategies of the other player. This distribution is adapted by Bayesian updating using the observed actions of the opponent. Each player uses the strategy which maximizes his expected payoff under the anticipated distribution of the others' mixed strategy. Eichberger [58] has shown that whenever the actually played strategies caused by Bayesian learning converge towards a pair of pure strategies, this pair of strategies has to be a Nash equilibrium, if all a priori distributions about the opponent's mixed strategy at time zero have full support. The same author has also shown that under some conditions on the initial a priori distributions the limit of the expected opponent's strategies has to be a Nash equilibrium, if a limit exists. The results about fictitious play can be seen as a special case of these general results, because the evolution of fictitious play equals Bayesian learning for the case where the a priori distributions about the opponents strategy are multivariate beta distributions (the family of these distributions is conjugated to the multinomial distribution).

Whereas the concepts of Bayesian learning assume that a game is played iteratively between two single players, there also exist a number of learning models in the evolutionary game theory literature, which are based on the presumption that the game is played within a population. Also in this case the learning behavior may be described by a best reply dynamics, as done in the well-received model of Kandori, Mailath and Rob [91]. They consider a 2×2

game played within a population where every individual in the population reviews his pure strategy every period with a given probability. Whenever he reviews his strategy he adopts the pure strategy which is a best reply to the current population distribution. The process is disturbed by errors or mutations, which occur with a small probability and lead to a random changing of the strategy. The process may be called a disturbed "Darwinian" dynamics, since the underlying deterministic dynamics has the so called "Darwinian" property, which says that the number of individuals playing a pure strategy increases if and only if the current payoff of this strategy is higher than the payoff of the other strategy. They analyze the limit of the stationary distribution of the resulting Markov chain, if the mutation probability tends towards zero. They show that the support of this limit distribution consists only of one state, which has to be a Nash equilibrium. In the case of coordination games, where two pure and one mixed equilibrium exist, the process will in the long run be concentrated on the risk dominant pure equilibrium. A pure equilibrium (s, s) is said to risk dominate another pure equilibrium (r, r) if the loss of deviation from s to r is larger than the loss of deviation from r to s (i.e. $a_{ss} - a_{rs} > a_{rr} - a_{sr}$). See Harsanyi and Selten [79] for a detailed discussion of the notion of risk dominance. Results for the same learning process in $n \times n$ games were subsequently derived in Kandori and Rob [92].

A different approach to model learning within a population was taken by Vega-Redondo [171]. He assumes that the individuals do not know the payoff function, and are therefore not able to play best replies. Instead he assumes that an individual reviewing his strategy will imitate one of the individuals which got the highest payoff in the last period. Additionally, it is again assumed that mutations occur with a small probability. We see that this model describes a situation which is rather similar to the setup we used to interpret a genetic algorithm as an economic learning process. Vega-Redondo analyzes the limit of the stationary distribution for small mutation probabilities, but only for a special class of games, namely the games of common interest[3]. He shows that for large populations his imitation process will end up with probability one in the Pareto dominant equilibrium. The same author studies a similar learning process in the context of an oligopoly and shows that for small mutation probabilities the unique long run outcome of the process has all firms choosing an output that maximizes profits when taking the market clearing price as given (Vega-Redondo [173]).

In Dawid [45] a population is considered where the individuals do neither know the payoff function of the game nor have enough information to imitate successful strategies. Some kind of satisficing approach is used to argue that in such a setup individuals change their strategies randomly from time to time, the more often the smaller their payoff is compared with the maximal obtainable payoff. It is shown that in this model an equilibrium will be

[3] A game is called a game of common interest, if there exists a Pareto dominant equilibrium in pure strategies.

reached with probability one if the number of strategies is not larger than three.

The standard approach to model learning in a long term evolutionary sense is the well known replicator dynamics (see Hofbauer & Sigmund [86] or Cressman [40] for an introduction). This dynamics was derived from standard assumptions about natural evolutionary processes, and models the impact of natural selection. The assumptions are basically that the evolutionary game is played within an infinite population, and that the number of offspring of an individual is proportional to its fitness. If m_t denotes the population distribution at time t, the dynamics is described by the following difference equation:

$$m_{it+1} = m_{it} \frac{e_i^T A m_t}{{m_t}^T A m_t} \tag{5.2}$$

This system of nonlinear difference equations is rather hard to analyze. Easy to see is however that every Nash equilibrium of a game is a fixed point of the replicator dynamics (5.2). It has also been shown that there is some connection between the notion of ESS, symmetric Nash equilibrium and the stability of the replicator dynamics (see Van Damme [169]). Every fixed point of (5.2) which is locally asymptotically stable has to be a symmetric Nash equilibrium. On the other hand every ESS is locally asymptotically stable with respect to the continuous replicator dynamics, but not necessarily with respect to the discrete time replicator dynamics. It is only possible in a few special games to completely analyze the stability properties of the equilibria. The reason for this is that in general the finding of suitable Liapunov functions is quite demanding for these dynamics. Unfortunately there exists no general technique to construct such functions. One of the cases where a Liapunov function has been found is that of the rock-scissors-paper game, which will be presented in section 5.3.1.

As already mentioned in chapter 4, the selection operator of a genetic algorithm with proportional selection has the form of the right hand side of the replicator dynamics. Of course, this is by no means surprising, because proportional selection incorporates exactly the properties which are assumed for deriving the replicator dynamics. Due to this connection between the replicator dynamics and the GA, one point of the following analysis is to show whether the behavior of these two learning dynamics is similar in all circumstances.

5.3 Learning by a Simple Genetic Algorithm

I use a simple genetic algorithm to model a population whose inhabitants play an evolutionary game against each other[4]. Every string represents one

[4] Some of the simulation results shown here were obtained in joint work with A. Mehlmann [50].

individual and encodes the strategy of this individual. The three standard operators are used to build up a new population each period. As we have seen in chapters 3 and 4, the encoding of the strategies may be very important for the learning behavior of the GA. In proposition 4.2.1 we have proven that in the long run the process will be in uniform states for most of the time. This means that the assumption that the individuals play only pure strategies would with a high probability rule out a priori the learning of mixed strategies. As this could be a harsh technical restriction, which would in many games rule out a priori the learning of any equilibrium, mixed rather than pure strategies are encoded. The assumption that the individuals play mixed strategies is by no means implausible and has already been used in evolutionary game theoretic models (e.g. Samuelson [149]).

In the following analysis always 3×3 evolutionary games will be considered. Every string encodes a mixed strategy in the three-dimensional simplex. Strings of length $l = 12$ represent the mixed strategies using the following coding technique. Every four bits encode an integer in the binary alphabet. This gives three integers i_1, i_2, i_3. The mixed strategy m, which is encoded by this string is now given by $m = \frac{1}{i_1+i_2+i_3+0.003}(i_1+0.001, i_2+0.001, i_3+0.001)$. The addition of the small terms should rule out the division by zero in the case of a string 0. The mixed strategy encoded by the binary string k is denoted by $m(k)$. The fitness of the string k is given by the payoff this string receives playing against the population distribution. Therefore, we get

$$f_k(\phi) = \sum_{j=0}^{r-1} \phi_j m(k)^T Am(j).$$

Setting $\phi = e_j$ we have $f_k(e_j) = m(k)^T Am(j)$. Now assume that the game defined by A has a strict symmetric Nash equilibrium in pure strategies, and that this pure strategy is coded by the string k. This implies that $f_k(e_k) > f_j(e_k)\ \forall j \neq k$. However, if k codes a mixed symmetric Nash equilibrium, at least two pure strategies are best responses, and, of course, also all mixed strategies consisting only of these two pure strategies. This means that there exist several strings j coding these strategies with $f_j(e_k) = f_k(e_k)$. From proposition 4.6.4 we know that a uniform state consisting only of strings which are a strict best reply to themselves will be locally attractive for a GA with a large population, regardless to the kind of crossover being used. Uniform states with strings which are no strict best replies to themselves may however be unstable. This implies that in our case a pure Nash equilibrium, where the equilibrium strategy is a strict best reply to itself (note that this implies evolutionary stability but not vice versa) is locally asymptotically stable for a GA with a large population. A mixed equilibrium may however be unstable. In one of the simulations we will see that a mixed equilibrium may be unstable, even if it is an ESS.

5.3.1 Rock-Scissors-Paper Games

The first class of games I use to analyze the learning abilities of a GA is the class of circulant Rock-Scissors-Paper (RSP) games. A game is called a circulant RSP game if its payoff matrix has the following form:

$$A = \begin{pmatrix} x & y & z \\ z & x & y \\ y & z & x \end{pmatrix} \qquad y > x \geq z. \tag{5.3}$$

It can be easily seen that the unique Nash equilibrium of such a game is given by $m^* = \frac{1}{3}(1, 1, 1)$. Further it can be shown that m^* is evolutionary stable if and only if $x < \frac{y+z}{2}$ holds [5]. RSP games are of special interest because of their circulant best reply structure. Every strategy is a best response to one of the other strategies, namely, to the strategy which is no best reply to this strategy. This structure makes it difficult for a population to learn the unique equilibrium. On the other hand, the special structure of the payoff matrix facilitates sometimes the analysis of the learning process. Weissing [181] was able to find a Liapunov function for the replicator dynamics in RSP games. Using this function he has shown that the equilibrium is asymptotically stable with respect to the replicator dynamics (5.2) if and only if $x^2 < yz$ holds.

In the first simulations I consider a circulant RSP game with $(x, y, z) = (1.1, 4, 1)$ (I call this game RSP1). The GA has a population of size $n = 100$ and uses proportional selection, one-point crossover with $\chi = 0.8$ and mutations with $\mu = 0.001$. Figure 5.1 shows the evolving behavior of the average mixed strategy of P projected onto the $m_1 m_2$-plane. This representation is chosen rater than the more common one in the simplex, because it is easier to determine the exact strategy corresponding to a given point in this representation. Due to the chosen values in the payoff function, the equilibrium m^* is evolutionary stable and also asymptotically stable with respect to the replicator dynamics. Nevertheless, we can see clearly from figure 5.1 that the GA does not converge to a uniform state corresponding to the equilibrium. It does not seem that the GA converges to a uniform state at all, but the trajectory oscillates in an area around the equilibrium. This behavior is clearly different from the behavior of the replicator dynamics. Of course, the replicator dynamics incorporates no mutations.

Let us look therefore at another simulation, where the same payoff matrix and the same GA is used with the only exception that now $\mu = 0$. The trajectory of this simulation can be seen in figure 5.2. We see that also the process without mutations does not settle into the equilibrium state, but into a state relatively close to the equilibrium. Thus, the GA without any mutations behaves quite differently from the replicator dynamics. This different behavior might have two different reasons. First, the replicator dynamics does not incorporate crossover effects, and second, the population is finite in

[5] See Appendix B.1 for the exact calculations.

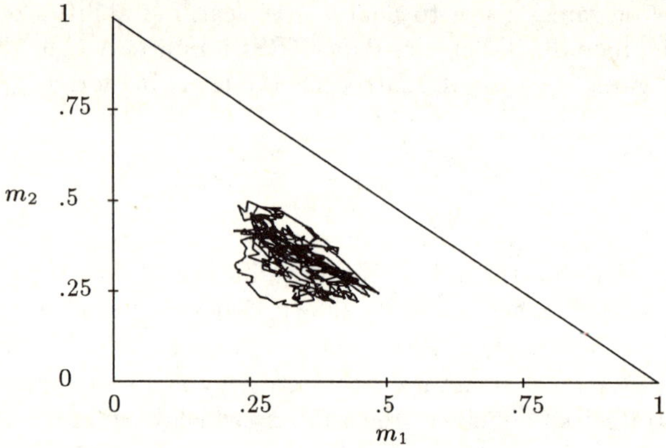

Fig. 5.1. The average population strategy of a SGA in a RSP game with the parameter values $(x, y, z) = (1.1, 4, 1)$, $n = 100$, $\chi = 0.8$ and $\mu = 0.001$

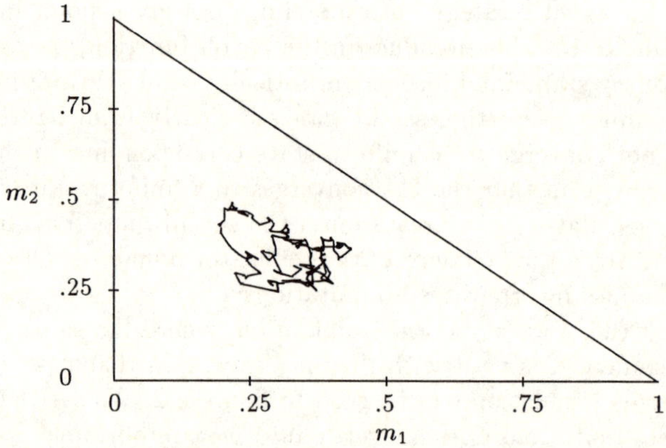

Fig. 5.2. The average population strategy of a SGA in a RSP game with the parameter values $(x, y, z) = (1.1, 4, 1)$, $n = 100$, $\chi = 0.8$ and $\mu = 0$

our simulations. In cases without mutations the variance within the population decreases steadily, for the selection operator reduces the variance, and the crossover operator keeps the variance on average but does not increase it. Therefore, the process settles down at some random point. If mutations increase the variance of the population in every period, the GA will not converge, but keep evolving, without reaching the equilibrium. Small mutations may be seen as a kind of compensation for the finiteness of the population, and this kind of trajectory therefore seems to be best suited to be compared with the replicator dynamics. Figure 5.2 shows that although the equilibrium state is not stable in a mathematical sense the trajectory at least stays in some neighborhood of the equilibrium.

To understand why the GA does not converge to the equilibrium although the replicator dynamics does, we also have to remember that in our setup every individual plays a mixed strategy. Thus, there is a high number of population states which yield the unique equilibrium strategy as population distribution and all but one of these states are not uniform. If the population is now near such a heterogeneous state, the selection pressure towards the equilibrium state is quite small, but, due to the fact that the population is heterogeneous, crossover disrupts the state quite a lot and the population strategy is randomly pushed around the equilibrium. This kind of behavior can be seen quite clearly in figure 5.1.

Considering these arguments we might anticipate that the equilibrium will get more and more repulsive the weaker the selection force towards the equilibrium is. We check this by two more simulations, where the parameter x is increased. In figure 5.3 the trajectory of the average population strategy for the parameter values $(x, y, z) = (2.4, 4, 1)$ (RSP2), $n = 100$, $\chi = 0.8$ and $\mu = 0.001$ is shown. In this case the equilibrium m^* is still ESS, but no longer stable with respect to the replicator dynamics. It can be seen clearly that the equilibrium is now really repulsive for the GA. A circling behavior of the GA can be clearly observed now, but in contrast to the replicator dynamics the cycle is in the interior of the simplex. The repulsive effect of the equilibrium gets even larger, if we further increase x to $x = 3.5$ (RSP3). The equilibrium m^* is now neither ESS nor stable for the replicator dynamics. The trajectory of the GA ends up circling near the boundary of the simplex. Thus, it seems that the behavior of the GA is closely connected to the behavior of the replicator dynamics, but it is only of minor importance whether an equilibrium is ESS.

Summarizing, we may say that in the case of a game with a circulant best reply structure the GA never learns the equilibrium. The limit points of the process are random but it depends on the sign of the expression $x^2 - \frac{yz}{2}$ whether the GA stays at least close to m^* or tends towards the boundary. These results are in complete accordance with the results from chapter 4. Note that every mixed strategy is a best reply to m^*, which implies that the linearization of (4.3) has the eigenvalue 1 with multiplicity l. As an eigenvalue

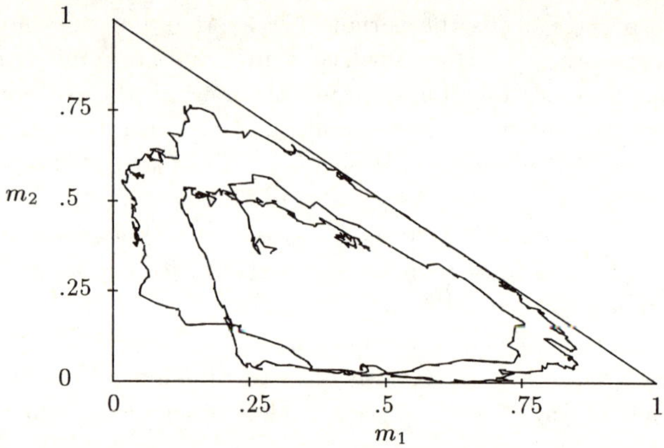

Fig. 5.3. The average population strategy of a SGA in a RSP game with the parameter values $(x, y, z) = (2.4, 4, 1)$, $n = 100$, $\chi = 0.8$ and $\mu = 0.001$

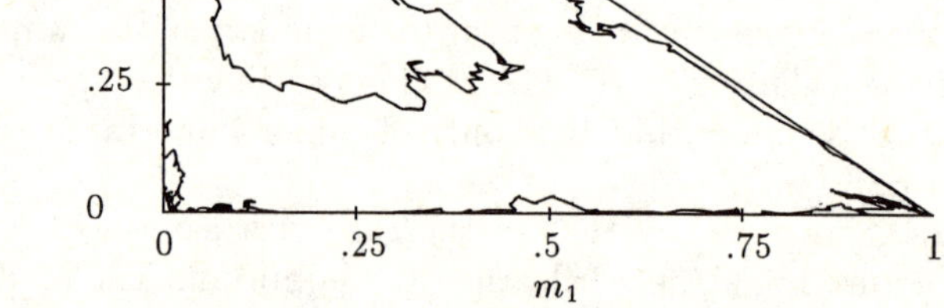

Fig. 5.4. The average population strategy of a SGA in a RSP game with the parameter values $(x, y, z) = (3.5, 4, 1)$, $n = 100$, $\chi = 0.8$ and $\mu = 0.001$

1 in a nonlinear system of difference equations allows no statement about the stability, a variety of different patterns of behavior are possible. We have seen that the crossover operator which has according to proposition 4.6.2 a stabilizing effect for uniform states has a destabilizing effect for heterogeneous equilibrium points. To speak in economic terms, information exchange leads to uniformity of behavior.

5.3.2 A GA Deceptive Game

The last example showed the problems of a GA in games with a circulant best reply structure. It is well known that the attaining of the equilibrium is hard for any learning process in these games. Also in the next example the GA will have problems to find the equilibrium, but here these problems would probably not occur with other learning rules as they are due to the finite population size of the GA. Consider an evolutionary game with the following payoff matrix

$$A = \begin{pmatrix} 2 & 0 & 1.5 \\ 0 & 6 & 1 \\ 0 & 36 & 0 \end{pmatrix}. \tag{5.4}$$

Let us call this game GAD. The only Nash equilibrium in this game is $((1,0,0),\ (1,0,0))$, and it is easy to see that this equilibrium is also ESS.[6] Note also that $(1,0,0)$ is the unique best reply to $(1,0,0)$, which implies by proposition 4.6.4 that the unique equilibrium state is locally asymptotically stable with respect to (4.3). Simulations show also that the replicator dynamics always converges towards the unique ESS $(1,0,0)$. Accordingly, one could anticipate that it must be easy for a GA to learn this equilibrium. Figure 5.5 shows the trajectory of a GA without mutations in this game. Contrary to our positive expectations the process again does not converge to the equilibrium, but stops somewhere on the line between $(0,0,1)$ and $(1,0,0)$. Considering the special structure of the payoff matrix of the game this is by no means a surprise. Due to the random initialization of the GA,the first generations' average strategy should be in the vicinity of $\frac{1}{3}(1,1,1)$. In such a situation the strings which put more weight on the third pure strategy are selected much more often than the rest. In particular this leads to the extinction of strings which favor the first pure strategy. However, as the population approaches $(0,0,1)$, the first pure strategy becomes the fittest. At this point in time the strategies preferring e_1 have been completely "forgotten", and the GA, being unable to converge to the ESS, stops half way through. Note that this simulation does not contradict proposition 4.6.2 as this proposition implies only the local stability of $(1,0,0)$. With random initialization the trajectory does in general not even come close to the equilibrium. From an economic point of view we might say that in the population there is too much imitation and

[6] See Appendix B.2.

too little innovation such that an early lock-in to a state may occur where some agents do not act optimally.

The situation resembles the case of "GA deceptive" functions. A function is called "GA deceptive", if the optimal solution includes some low order schemata whose average fitness is low. The GA tends to settle down at some local maximum, for the early building blocks that spread in the population are no realizations of these optimal schemata (see Goldberg [73]). In our game the same applies not to blocks but to whole strings. This is of course due to fact that, contrary to standard GA applications, the fitness function changes as the population evolves. I call games like GAD, where the early selection process leads to the extinction of equilibrium strategies, "GA deceptive games".

There are several ways to circumvent the problems of genetic algorithms in GA deceptive games. We could increase the size of the population, which would result in an increase of the length of memory and lead in the limit to a situation modelled by the replicator dynamics, where no strategy will ever be forgotten. The standard approach in evolutionary computation would of course be to add mutations to the process. In our interpretation this means, that the agents are able to reinvent the equilibrium strategies, even if they have forgotten them before. If we set the mutation probability μ to 0.001, the GA escapes the area around e_3 and converges to the ESS (see figure 5.6). The population finds the equilibrium, but due to the mutations, the uniform state is disrupted again and again, which causes the small fluctuations in the average population strategy in figure 5.6.

Another approach could be to prevent the population from forgetting the equilibrium strategies. A way to do this is to insert conservative agents, which are not willing to adapt their strategies, into the population. Technically speaking we add a fixed number of strings coding each pure strategy to the population. These strings are transferred unchanged to the next generation, but, of course, they can be selected into the mating pool and therefore enter the "ordinary" adaptive population. This construction is supposed to describe the fact that although conservative agents themselves are not influenced by their surrounding, they are able to change the behavior of the adaptive agents. The approach to consider populations whose individuals show different learning behavior is not completely new. As already mentioned in subsection 2.4, Beltrati and Margarita [19] use similar ideas to simulate an artificial stock market. Another paper where individuals with heterogeneous adaptive behavior are considered is by Biccheri and Rovelli [25]. They use an evolutionary setup to investigate how cooperative norms can be established. In particular, they analyze the dynamics of corruption and consider two types of individuals, namely, the strategic players which adapt their strategies and the irreducible ones who don't. They observe that "*... a crucial role in the establishment of a new norm is played by a small percentage of irreducible honest individuals. Such individuals are not adaptive in the sense that they*

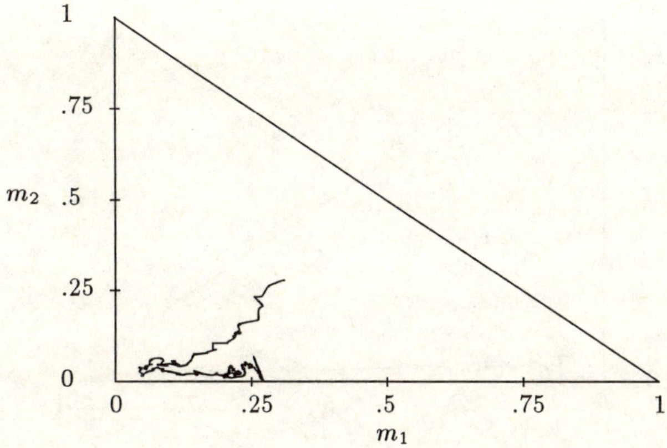

Fig. 5.5. The average population strategy of a SGA in the GA deceptive game GAD with the parameter values $n = 100$, $\chi = 0.8$ and $\mu = 0$

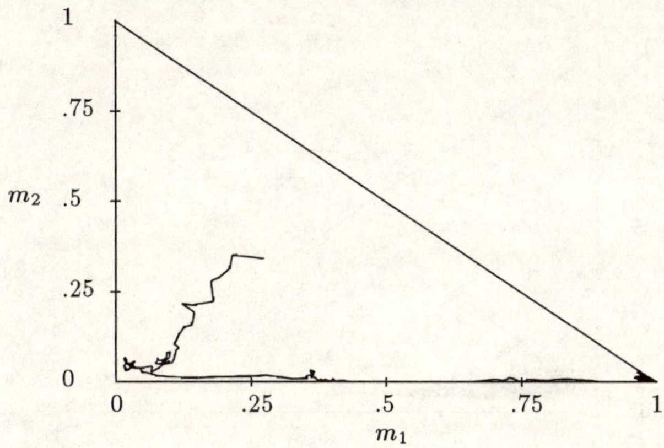

Fig. 5.6. The average population strategy of a SGA in the GA deceptive game GAD with the parameter values $n = 100$, $\chi = 0.8$ and $\mu = 0.001$

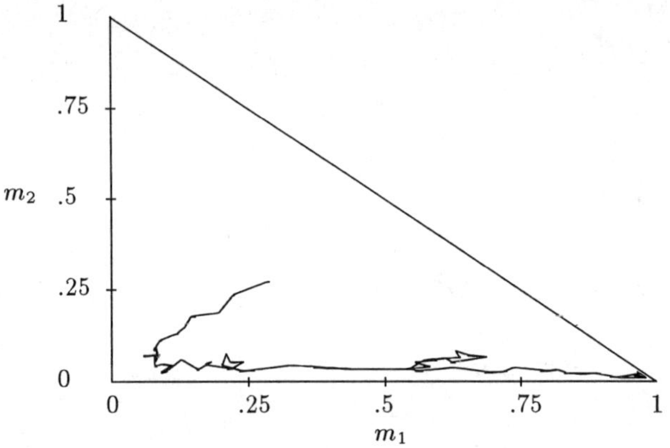

Fig. 5.7. The average population strategy of a SGA in the GA deceptive game GAD with the parameter values $n = 100$, $\chi = 0.8$, $\mu = 0$ and one conservative agent for each pure strategy

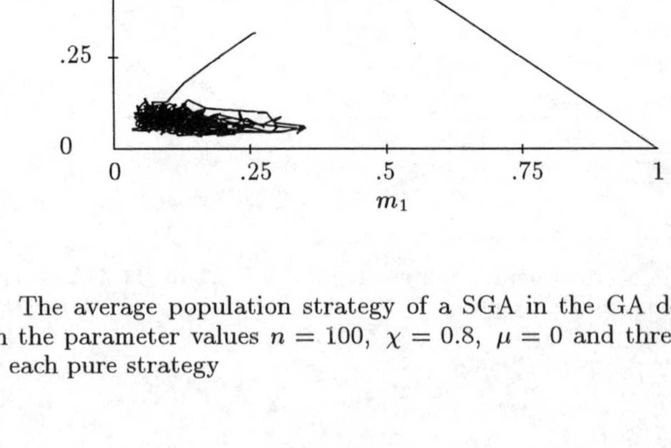

Fig. 5.8. The average population strategy of a SGA in the GA deceptive game GAD with the parameter values $n = 100$, $\chi = 0.8$, $\mu = 0$ and three conservative agents for each pure strategy

never change their strategy." Our conservative agents have quite the same properties as these irreducible individuals with the only difference that they may be imitated by the other adaptive individuals and may also influence them by information exchange.

The introduction of the conservatives bears two different effects. On one hand, no strategy can be completely extinguished, which means that due to selection and crossover any strategy can be reached at any time. On the other hand, conservative agents may prevent the population state from reaching an economic equilibrium state. If the number of conservatives is sufficiently large the non-adaptive agents could be exploited by the rest of the population, leading to off-equilibrium outcomes. Both effects can be seen very clearly in GAD: First, one conservative agent for each pure strategy is inserted into the population of 100 agents. This is enough to allow the GA to find the ESS, but the effect on the selection process is negligible (see figure 5.7). The learning of the equilibrium is even better than in the case with mutations, in the sense that the fluctuations of the average population strategy around $(1,0,0)$ are smaller. However, if 6 more conservative agents are introduced, the second of the two effects takes over and traps the GA in the area near e_3 (see figure 5.8). A small number of agents playing e_2 is exploited by the rest of the population and, as they are not willing to change their low fitness strategy, the equilibrium can not be reached.

The implications of these considerations are obvious. A small number of non-adaptive strings in a population is useful as some kind of constant memory, but the number of these strings has to be kept small enough, so that they do not affect the average population strategy significantly. However, we have to be aware of the fact that the facilitating of learning by non adaptive individuals only works in a "small world". If the number of pure strategies is very large the non adaptive individuals have to be a rather large fraction of the population, which of course weakens the ability of the whole population to evolve. If we only consider a small system, we may draw the conclusion that the rigidity of certain individuals may become useful for the whole population.

5.3.3 Learning in Non Deceptive Games

The last two examples have exposed the inability of the GA to learn equilibria in certain difficult circumstances. However, this is not a general phenomenon. In most cases, where no circulant or misleading best reply structure occurs, the GA has no problems in reaching one of the Nash equilibria of the game. To illustrate this remark let us consider now an evolutionary game with two equilibria in pure strategies, which has no misleading structure. This game called $G1$ has the payoff matrix

$$A = \begin{pmatrix} 4 & 2 & 2 \\ 2 & 1 & 2 \\ 1 & 1 & 6 \end{pmatrix}. \tag{5.5}$$

The game has two equilibria in pure strategies, $((1,0,0),(1,0,0))$ and $((0,0,1), (0,0,1))$, which are both ESS, and one symmetric mixed equilibrium, namely, $\frac{1}{7}(4,0,3)$[7]. It is easy to see that $((0,0,1),(0,0,1))$ risk dominates $((1,0,0),(1,0,0))$ and as can be seen in figure 5.9, the average population strategy of a population in a GA converges towards the risk dominant equilibrium. However, the convergence against the risk dominant strategy is not a necessary property of the learning behavior of a GA. The concept of risk dominance takes into consideration only the payoffs, which the different equilibrium strategies earn against each other, but not the payoffs against strategies which do not occur in either of the equilibria. At least in the beginning of the run, however, there will be strings in the population which attain positive probabilities to such pure strategies.

Following this line of thought let us increase the payoff of the first pure strategy against the second pure strategy. This will shift the direction of the initial selection more and more towards $(1,0,0)$. To get a stable convergence against $((1,0,0),(1,0,0))$ with an overwhelming probability we have to increase this value up to 6. The payoff matrix is now given by

$$A = \begin{pmatrix} 4\,6\,2 \\ 2\,1\,2 \\ 1\,1\,6 \end{pmatrix},$$

and this new game is called G2. The equilibria of G2 are just the same as in G1, and (e_1, e_1) of course still risk dominates (e_3, e_3). However the average population strategy of the GA approaches $(1,0,0)$ in G2, as can be seen in figure 5.10. This shows that a static concept like risk dominance, which compares only the payoffs of equilibrium states, is in some cases ill-suited to predict the outcome of an evolutionary process.

I have checked in several different evolutionary games that the GA in general reaches one of the ESSs, or at least one of the Nash equilibria. Contrary to the counterexamples presented in section 5.3.1 and 5.3.2, most of these games have no particularly interesting structure, and thus I have decided to present here only two examples of the many games where the GA finds the equilibrium, namely G1 and G2. Summarizing, we might say that in general genetic learning finds a Nash equilibrium in an evolutionary game. Problems may occur due to the finite size of the population in games where the strings with a high fitness in the beginning are no part of the equilibrium, or due to some kind of overshooting effects in games with a circulant best reply structure. These results are quite satisfactory, because the GA does a good job in simple straightforward cases, and has problems in cases where we would expect them also from an economic point of view. All the simulation results are in accordance with the theoretical findings in chapter 4, and in the case of the GA deceptive games we have used one of the useful notions of a GA mentioned in chapter 2. By inserting the conservative individuals we

[7] See Appendix B.3.

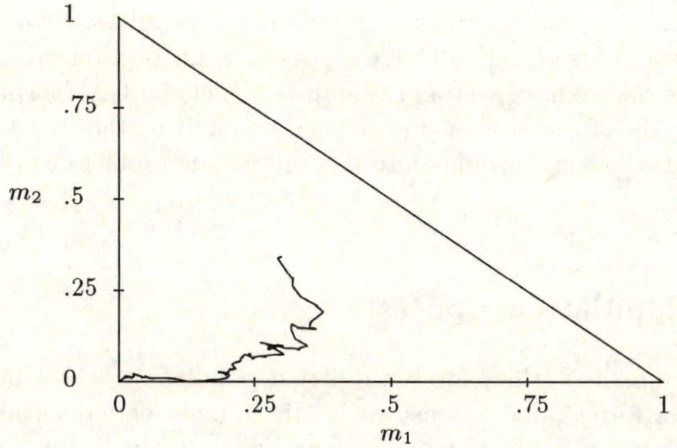

Fig. 5.9. The average population strategy converging towards the risk dominant equilibrium in G1 ($n = 100$, $\chi = 0.8$, $\mu = 0$)

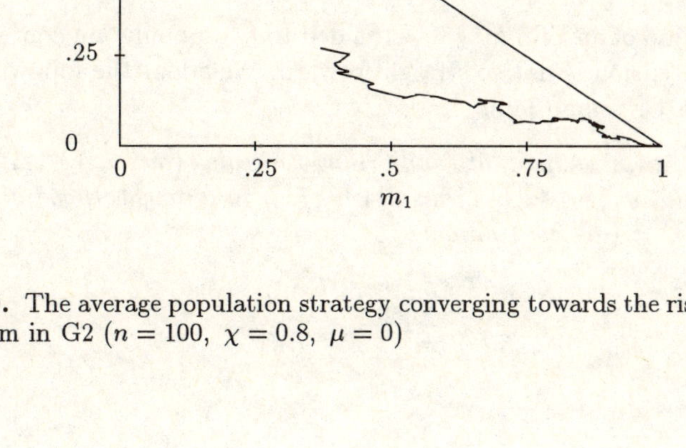

Fig. 5.10. The average population strategy converging towards the risk dominated equilibrium in G2 ($n = 100$, $\chi = 0.8$, $\mu = 0$)

have created a heterogeneous population, where not all individuals adapt to the changing surrounding in the same manner. It would be quite interesting to extend these kind of algorithms by creating a population containing of several different kinds of individuals with different adaptation rates or adaptation policies. As we have seen in the simple model of a GA deceptive game the limit outcome of the process may depend crucially on the structure of the population. Maybe some insights into this interesting problem can be gained in future work.

5.4 Two-Population Contests

Now let us assume that there are two different populations whose individuals play a normal form game against each other. Each period each agent in population one is randomly matched with several individuals from population two and plays the game with each of them. The same holds true for each individual in population two and, again, the number of individuals each agent is matched with is sufficiently large to claim that his fitness is equal to his expected payoff.

Two species models have already been dealt with in the evolutionary games literature, but often also intraspecific contests are allowed (see Cressman [40]). In our setup an individual never plays a game against another individual of the same species. Thus we consider a situation where a bimatrix game (A, B) is played between the two populations. Denote the number of pure strategies which may be used by an individual in population i by I^i and $|I^i|$ by N^i. As we are no longer restricted to symmetric strategy pairs we get the following standard definition of a Nash equilibrium.

Definition 5.4.1. *A pair of mixed strategies* $(m^1, m^2) \in \Delta^{N^1} \times \Delta^{N^2}$ *is called a Nash equilibrium of the bimatrix game* (A, B) *iff*

$$m^{1T} A m^2 \geq \tilde{m}^{1T} A m^2 \quad \forall \tilde{m}^1 \in \Delta^{N^1}$$
$$m^{1T} B m^2 \geq m^{1T} B \tilde{m}^2 \quad \forall \tilde{m}^2 \in \Delta^{N^2}.$$

Also the notion of an ESS may be extended to two-population contests. However, this extension is not so straightforward. We adopt the following definition given in Cressman [40]:

Definition 5.4.2. *A pair of equilibrium strategies* $(m^1, m^2) \in \Delta^{N^1} \times \Delta^{N^2}$ *is called evolutionary stable if for all* $(\tilde{m}^1, \tilde{m}^2)$ *in a neighborhood of* (m^1, m^2) *with* $(\tilde{m}^1, \tilde{m}^2) \neq (m^1, m^2)$ *either*

$$m^{1T} A \tilde{m}^2 > \tilde{m}^{1T} A \tilde{m}^2 \quad or$$
$$\tilde{m}^{1T} B m^2 > \tilde{m}^{1T} B \tilde{m}^2$$

holds.

Using this definition we have the following important result which is due to Selten [155]

Theorem 5.4.1. *Let* $(m^1, m^2) \in \Delta^{N^1} \times \Delta^{N^2}$ *be an ESS of the bimatrix game* (A, B) *then* m^1 *and* m^2 *are pure strategies.*

This result may suggest that in two-population contests mixed equilibria are hard to learn by an evolutionary adaptation process. In this section we will see that for GAs this indeed holds true.

If we assume that the state of the two populations is updated by using two GAs we have an example of a two-population model in the sense of section 4.7. The two fitness functions are given by:

$$f_j^1(\phi^1, \phi^2) = \sum_{i=0}^{r^2-1} \phi_i^2 m^1(j)^T A m^2(i) \quad j \in \Omega^1$$
$$f_k^2(\phi^1, \phi^2) = \sum_{i=0}^{r^1-1} \phi_i^1 m^1(i)^T B m^2(k) \quad k \in \Omega^2,$$

where $m^i(j)$ denotes the mixed strategy in Δ^{N^i} which is encoded by $j \in \Omega^i$. Note that $f^1(\phi^1, \phi^2)$ depends only on ϕ^2 and $f^2(\phi^1, \phi^2)$ depends only on ϕ^1. According to our analytical results any strict economic equilibrium of the system is locally asymptotically stable with respect to the expected dynamics of the GA but also non equilibrium states may be stable. It can be easily seen that in this setup a state of the system corresponds to a strict economic equilibrium as defined in definition 4.7.1 if and only if the corresponding strategies are evolutionary stable. Thus, evolutionary stable strategies are always stable fixed points of the dynamical system but evolutionary stability is no necessary property for a state to be stable. This discrepancy originates from the use of the crossover operator which may make evolutionary unstable states stable.

The games considered here are all symmetric games where $I^1 = I^2$ and $B = A^T$. We will look at some two populations simulations in games which have already been used in the previous sections. The coding of the strategies in our GA is exactly the same as in the one population simulations and again only the three standard operators are applied. Both populations have the same size, namely $n^1 = n^2 = 100$ and the fitness functions are given by the expressions stated above.

First, let us again look at a circulant Rock-Scissors-Paper game, where the payoff of the two players is given by (A, A^T) with A given by (5.3). Note that the strategy profile $(\frac{1}{3}(1,1,1), \frac{1}{3}(1,1,1))$ is still a Nash equilibrium of the game but it can never be evolutionary stable in this two-population context. In figure 5.11 the trajectories of the average population strategies of the two populations in a simulation with parameter values $(x, y, z) = (1.1, 4, 1)$, $\chi = 0.8$ and $\mu = 0.001$ are depicted.

Recall that for these values of x, y and z the central point m^* is evolutionary stable in the one population context and the trajectory of a one population GA stays close to this equilibrium. Quite contrary to this the

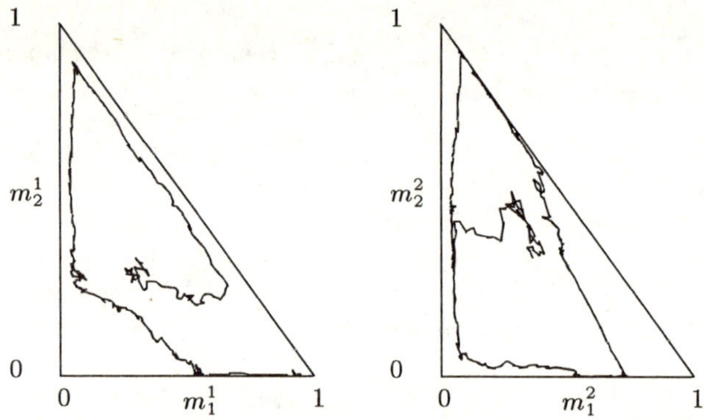

Fig. 5.11. The trajectories of the average population strategies in a two population SGA in a RSP game with parameters $(x, y, z) = (1.1, 4, 1), n^1 = n^2 = 100, \chi^1 = \chi^2 = 0.8, \mu^1 = \mu^2 = 0.001$

trajectories of the average population strategies in the two-population contest move away from the central point and approach the boundary of the simplex. Similar behavior was observed in all simulations carried out in RSP games. We conclude from these results that the inner equilibrium of the game is always unstable in two-population contests. Note that our analytical results from section 4.7 allow no statement about the stability of the inner equilibrium as the largest eigenvalues of the linearization have modulus one.

Next, we deal with the behavior of two population GAs in GA deceptive games. Consider the bimatrix game (A, B) where $B = A^T$ and A is given by (5.4). We know already that the only Nash equilibrium in this game is given by $((1,0,0),(1,0,0))$ and it is easy to verify that this equilibrium is also evolutionary stable in the two population setup. Thus, we know from proposition 4.7.4 that this equilibrium is locally asymptotically stable with respect to the dynamics of the GA. In figure 5.12 we see the result of a simulation with a two population GA where the mutation probability is zero in both populations. Similar problems as in the case of a one population contest occur. Both trajectories head initially towards e_3. As soon as one of the two population strategies is sufficiently close to this vertex the best response of the individuals in the other population is to play their first pure strategy and this trajectory turns around. However, the genetic material which is needed to approach e_1 has been lost already and the trajectory stops near the center of the simplex. The population strategy of the other population stays close to e_3 as this is a best reply now. Thus, the population which gets close to e_3

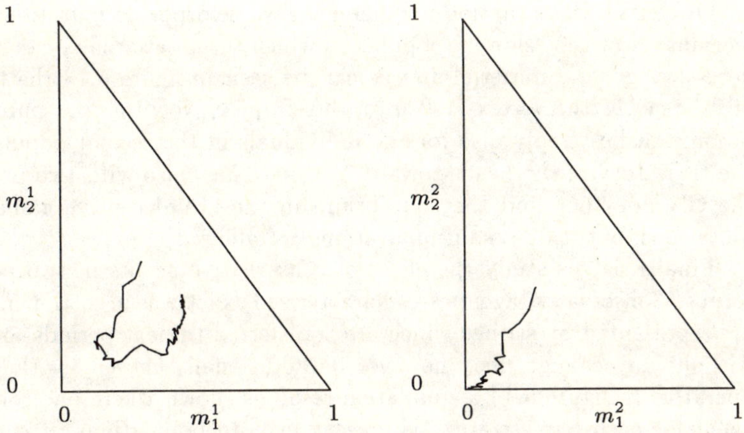

Fig. 5.12. The trajectories of the average population strategies in a two-population SGA in the GA deceptive game GAD with parameters $n^1 = n^2 = 100$, $\chi^1 = \chi^2 = 0.8$, $\mu^1 = \mu^2 = 0$

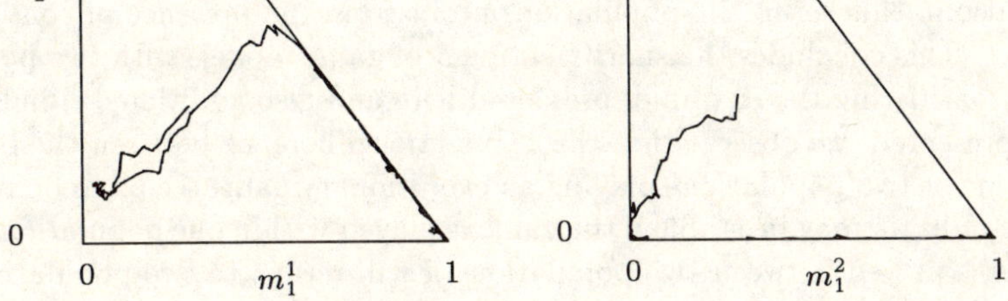

Fig. 5.13. The trajectories of the average population strategies in a two-population SGA in the GA deceptive game GAD with parameters $n^1 = n^2 = 100$, $\chi^1 = \chi^2 = 0.8$, $\mu^1 = \mu^2 = 0.001$

first will, in the long run, have a higher average payoff than the population which turns around.

Of course, the situation is different if we incorporate mutations into the learning. Now the "slower" population which turns towards e_1 can regain the necessary genetic material and as can be seen in figure 5.13 the trajectory of m^1 reaches the vertex $(1,0,0)$ of the simplex. Now the first pure strategy becomes a best reply also for the individuals in the second population and the trajectory of m^2 heads towards e_1 too. Thus, also with two populations the GA does not find the equilibrium in the GA deceptive game without mutations, but it does so if innovations are allowed.

Finally, let us study the effect of conservative agents in two population setups. Conservative agents are characterized exactly as in section 5.2.2. They are represented by strings which are transferred to next periods' population without any change, may however be additionally chosen by the selection operator. In figure 5.14 a simulation result is shown where one conservative agent for each pure strategy is present in both populations. It can be seen that the effect the conservative agents have is quite similar to that of the introduction of mutations. Both average population strategies end up near the equilibrium strategy $(1,0,0)$. The effect of a symmetric introduction of conservative agents is therefore quite similar to the effect they have in one population contests.

However, with two populations we may also consider an asymmetric scenario where only one of the two populations contains conservative agents. Such a situation where three conservative agents are present in the second population but none in the first one can be seen in figure 5.15. Due to the conservative agents the second population is able to approach the equilibrium strategy $(1,0,0)$ which is a best reply to the average population strategy of population one. The strategy of the first population on the other hand can not recover the strategies which put a high weight on the first pure strategy and gets stuck near e_3. Whereas the average payoff of the individuals in the first population is 1.01 the three conservatives in the second population make it possible that the average payoff in their population is 1.55. We see that also in this setup the populations profit from the presence of conservatives.

This concludes the short treatment of game models with two populations. Considering the examples presented here and also additional simulations not presented, we observe that the only major difference between the behavior of one or two populations playing an evolutionary game seems to be that mixed equilibria may be stable if the game is played within one population but not if it is played between two populations. I will return to two population models with a different structure in the next chapter.

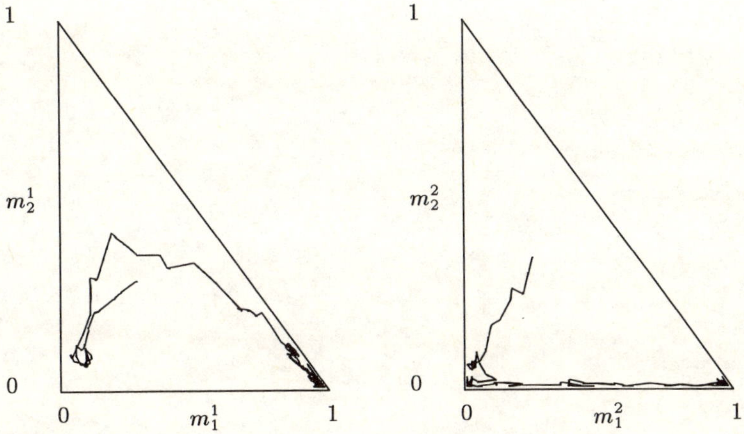

Fig. 5.14. The trajectories of the average population strategies in a two-population SGA in the GA deceptive game GAD with parameters $n^1 = n^2 = 100$, $\chi^1 = \chi^2 = 0.8$, $\mu^1 = \mu^2 = 0$ and one conservative agent for each pure strategy in both populations

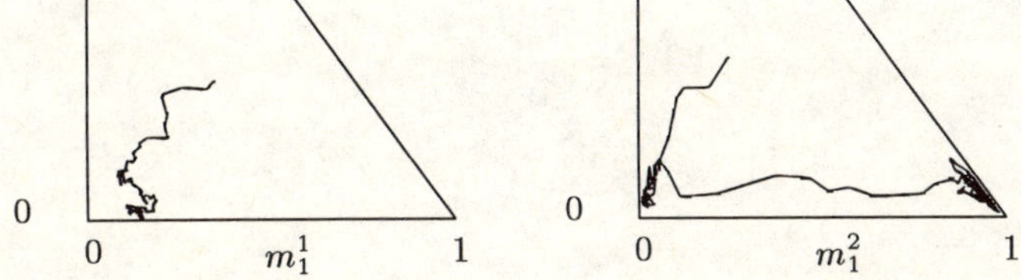

Fig. 5.15. The trajectories of the average population strategies in a two-population SGA in the GA deceptive game GAD with parameters $n^1 = n^2 = 100$, $\chi^1 = \chi^2 = 0.8$, $\mu^1 = \mu^2 = 0$ and one conservative agent for each pure strategy only in the second population

6. Simulations with Genetic Algorithms in Economic Systems

In chapter 5 I presented some simulations of genetic learning in evolutionary games. Evolutionary games, however, are always a stylized model of an economic system, since no explicit structure of the model is given. In many instances the special structure of a certain economic model allows considerably more precise insights into the behavior of individuals in such a situation than a more abstract game theoretic model. This holds also in particular true if we think of the adaptive behavior of boundedly rational agents. Basic dynamic properties of learning dynamics like convergence, dynamic equilibrium selection or complex behavior obviously depend on the framework the corresponding agents are living in. The purpose of this chapter is to study the dynamic properties of genetic learning in several standard economic models. Hereby we should not only try to derive these properties but also to understand their relation with the properties of the corresponding model. Altogether I will present three economic models. The first one is a very simple model of a market of price takers, where we will be able to see the power of the theoretical results from chapter 4 very impressively. The second model, an overlapping generations model with fiat money, has a more complicated setup, and does neither fit into the one nor the more population models analyzed in section 4. Therefore we have no theoretical results for this model, but we will observe in simulations that GAs are also able to learn cyclical and stochastic equilibria. The third model describes the interaction of buyers and sellers in sealed bid double auction market. This model formally fits into the two-population framework of section 4.7 and we will be able to rely on analytical and numerical insights in our analysis.

6.1 A Model of a Competitive Market

In this section I analyze a very simple model of a competitive market which is in the literature referred to as the cobweb model[1]. It describes the temporary equilibrium market prices in a single market with one lag in supply, i.e. since the production takes time, quantities produced must be decided before a market price is observed. The model was introduced by Leontief [108], who

[1] This section is based on a joint paper with M. Kopel [49].

postulated a linear model of demand and supply in which agents forget all except their most recent experiences. As is well-known, convergence to an equilibrium price occurs in this model for the cobweb stable case, namely if the supply is less elastic than the demand. The first dynamic analysis of the classic linear cobweb model when agents have memory was by Nerlove [132]. This was extended by Muth [125] to an analysis of rational learning with memory. Carlson [37] concluded that the linear cobweb model was stable when agents used the mean of past prices as their forecast price. The issue of the convergence of agents' learning towards a rational expectations equilibrium has also been addressed in a cobweb model. Bray [31] and Bray and Savin [33] demonstrated that, if agents use a least-squares learning procedure, prices in a cobweb model almost surely converge to a rational expectations equilibrium. Contrary to their "ad hoc" learning scheme, Bray and Kreps [32] demonstrated that in the context of this model there is a unique equilibrium with rational learning. Besides these analytical examinations of adaptive behavior in the cobweb model there are several papers where artificially intelligent agents are considered. The papers by Arifovic [8] and Franke [64] mentioned in section 2.4.3 fall into this category. The model considered here differs slightly form the standard model because it is assumed that a firm has short term fixed costs of production, like labour costs or the costs of warehouse renting. Such models with fixed costs have been recently analyzed in the literature, see for example Day [53].

There are n competitive firms which produce and offer the same consumer-good. Each firm f_i produces at every time t a quantity $y_{i,t}$ of this consumer-good. All firms have the same convex cost function, namely

$$c(y) = \begin{cases} \alpha + \beta y^2, & \alpha, \beta > 0 \; y > 0 \\ 0 & y = 0 \end{cases}.$$

The parameter α denotes the short term fixed costs of the firm, which only incur if the production quantity of the firm is positive. The supply of the consumer-good at time t is given by

$$s_t = \sum_{i=1}^{n} y_{i,t}.$$

We assume that the demand for the consumer-good, d_t, decreases linearly with the price of the good, i.e.

$$d_t = \max[\gamma - \delta p_t, 0], \quad \gamma, \delta > 0.$$

The market clearing condition implies that the price of the good at time t is given by the inverse demand function

$$p_t = \frac{\gamma}{\delta} - \frac{1}{\delta} \sum_{i=1}^{n} y_{i,t}.$$

In order to allow the clearance of the market $\sum_{i=1}^{n} y_{i,t} < \gamma$ has to be satisfied. To keep the notation simple we write this equation as

$$p_t = a - \frac{b}{n} \sum_{i=1}^{n} y_{i,t}. \tag{6.1}$$

The profit of a firm f_i is given by

$$\Pi(p_t, y_{i,t}) = p_t y_{i,t} - c(y_{i,t}) = \begin{cases} p_t y_{i,t} - \alpha - \beta y_{i,t}^2 & y_{i,t} > 0 \\ 0 & y_{i,t} = 0 \end{cases}.$$

In order to guarantee that the profit is positive at least for some values of p and y we have to assume

$$\alpha < \frac{a^2}{4\beta}. \tag{6.2}$$

In cases where the fixed costs are so high that (6.2) does not hold there would be no incentive for a firm to enter the market. Such a scenario is therefore not interesting from an economic point of view. The decision maker in firm i does not know the price p_t, when he has to decide which quantity to produce at time t. However, he has an expectation $p_{i,t}^e$ about the price at time t. Given this expectation the optimal quantity to produce is given by

$$y^*(p_{i,t}^e) = \begin{cases} \frac{p_{i,t}^e}{2\beta} & p_{i,t}^e > 2\sqrt{\alpha\beta} \\ \{0, \sqrt{\frac{\alpha}{\beta}}\} & p_{i,t}^e = 2\sqrt{\alpha\beta} \\ 0 & p_{i,t}^e < 2\sqrt{\alpha\beta} \end{cases} \tag{6.3}$$

We say that an individual has perfect foresight if his expectation about p_t coincides with the afterwards observed value of p_t for all t. In a homogeneous rational expectations equilibrium all firms have perfect foresight and all take the same optimal actions (i.e. $p_i^e = p^* \ \forall i = 1, \ldots, n$ and $y_i = y^* \ \forall i = 1 \ldots n$ for some $y^* \in y^*(p^*)$ with $p^* = a - by^*$). We have to consider two cases. First assume that the expected price is larger or equal than $2\sqrt{\alpha\beta}$. In this case, inserting (6.3) into (6.1) gives

$$p^* = a - by^*(p^*) = a - b\frac{p^*}{2\beta},$$

which yields as equilibrium price

$$p^* = \frac{2\beta a}{2\beta + b}. \tag{6.4}$$

This expression has to be larger or equal than $2\sqrt{\alpha\beta}$, and this holds if and only if

$$\alpha \leq \frac{\beta a^2}{(2\beta + b)^2}. \tag{6.5}$$

On the other hand, if we assume that the rational equilibrium price is smaller or equal than $2\sqrt{\alpha\beta}$, we get from (6.3)

$$p^* = a - by^*(p^*) = a.$$

The equilibrium price

$$p^* = a$$

is smaller or equal than $2\sqrt{\alpha\beta}$ if and only if

$$\alpha \geq \frac{a^2}{4\beta}.$$

However such a parameter constellation contradicts (6.2). This leaves us with two possible scenarios. If (6.5) holds there is a unique rational expectations equilibrium with an equilibrium price p^* given by (6.4), otherwise no homogeneous rational expectations equilibrium exists in the model.

6.1.1 Pure Quantity Decision

To simulate the evolving behavior of the system described above, each firm is represented by a binary string in a genetic algorithm. Contrary to the next example the quantity the firm decides to produce is encoded rather than the price expectations of the firm. Every binary string b_i in the population P_t encodes a real number in $[0, \frac{a}{b}]$, namely the quantity the firm f_i decides to produce at time t. The price at time t is calculated by using (6.1). The profit of a firm may become negative for too small prices. To rule out negative fitness values for single strings, define the fitness function as

$$f_k(\phi) = \Pi(p(\phi), y(k)) + C,$$

where $y(k)$ denotes the production quantity encoded by the string $k \in \Omega$, $p(\phi)$ the price which emerges from (6.1), if the whole population is in state $\phi \in S$ and $C = \alpha + \beta\left(\frac{a}{b}\right)^2$.

In the first set of simulations the following parameter values are used:

$$a = 5, \quad b = 5, \quad \alpha = 0.25, \quad \beta = 1. \tag{6.6}$$

It is easy to check that the conditions (6.2) and (6.5) are satisfied for these parameter values, which implies that a unique rational expectations equilibrium exists with an equilibrium output of

$$y^* = y^*(p^*) = \frac{a}{2\beta + b} = \frac{5}{7}.$$

As in the simulations in evolutionary games a simple genetic algorithm in-

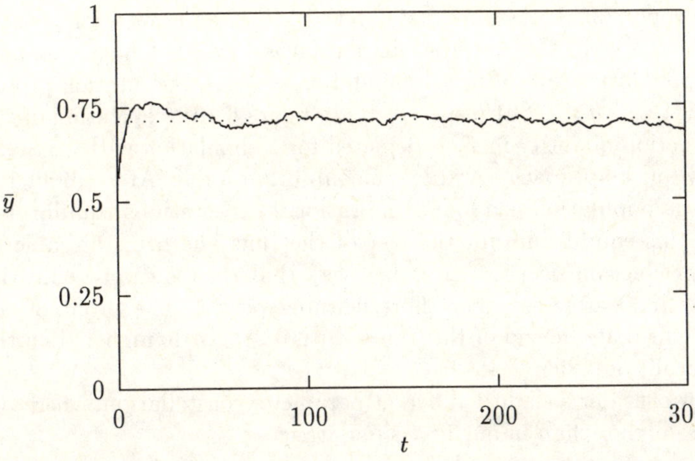

Fig. 6.1. The average production quantity in P_t for $a = 5$, $b = 5$, $\alpha = 0.25$ and $\beta = 1$. The dotted line describes the rational expectations equilibrium ($n = 1000$, $\chi = 1, \mu = 0.001$)

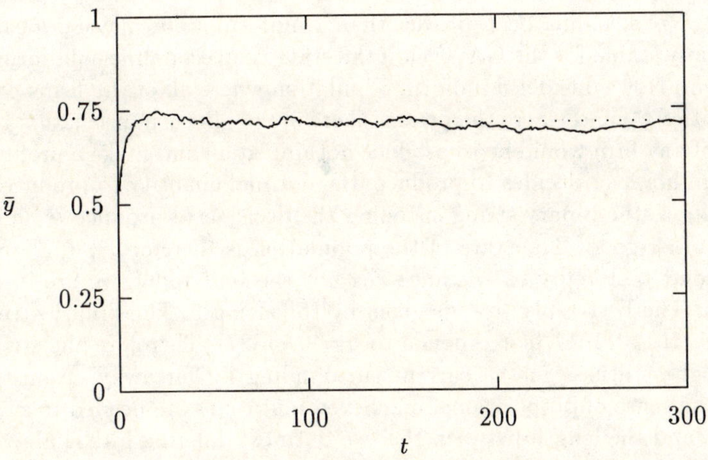

Fig. 6.2. The average production quantity in P_t for $a = 5$, $b = 5$, $\alpha = 1$ and $\beta = 1$. The dotted line describes the rational expectations equilibrium for the cases where $\alpha < \frac{\beta a^2}{(2\beta+b)^2}$ holds ($n = 1000$, $\chi = 1$, $\mu = 0.001$)

corporating proportional selection, one-point crossover and mutation is used. The length of the string in the first simulation is $l = 10$. I have chosen a rather large population size of $n = 1000$ and crossover and mutation probabilities of $\chi = 1$ and $\mu = 0.001$ respectively. In figure 6.1 the evolution of the average production quantity in P_t is depicted for a simulation with an SGA. We see a very quick approach towards the equilibrium value. After about 100 generations the population has found the rational expectations equilibrium, and stays in this equilibrium for the rest of the run. The firms are able to adapt their production decision in such a way that their output quantities are always optimal after a rather short learning period. The profit of the representative firm at the end of this run is with 0.28 a little higher than the equilibrium profit of 0.255.

Let us now consider a slightly different parameter constellation, where the fixed costs are higher than in our first simulation:

$$a = 5, \quad b = 5, \quad \alpha = 1, \quad \beta = 1. \tag{6.7}$$

Condition (6.2) is still satisfied, but (6.5) no longer holds. This implies that in this model no homogeneous rational expectations equilibria exist. Looking at the simulation in this model (figure 6.2) we observe however that the population of firms represented by the GA behaves exactly in the same way as in the model with a rational expectations equilibrium. The average produced quantity converges towards the value y^* and the price therefore towards p^* of (6.4). An explanation for this kind of behavior can be found in proposition 4.6.2, where it has been proven that a uniform state may be locally asymptotically stable for the GA, even if the state is no economic equilibrium of the system. Now consider a uniform population where all of the firms produce quantities y^*, and where the price is therefore p^*. In such a situation the best reply of any firm would be to produce nothing at all and make a profit of 0. If the firm however decides to produce, the optimal quantity to produce is y^*. Let i denote the binary string encoding the decision to produce y^* units of the consumer-good. The state of the population is therefore $e_i \in S$, and i is the second best reply to e_i. Since i is not the best reply, we are in no equilibrium. The best reply to e_i is given by the string 0. The binary string i is given by 1011011011 (corresponds to $y = 0.71387$). Therefore the string with the highest fitness in the current surrounding is "far away" from the string i in the sense of proposition 4.6.2. Seven mutations are needed to swap from i to 0, and the length between the two outmost differing bits is nine. A firm producing nothing at all would have a profit of 0, which would in the given environment yield a fitness of $f_0(e_i) = 2$, whereas the string i gets only $f_i(e_i) = 1.51$. Nevertheless, the condition of proposition 4.6.2 is satisfied (i.e. $\frac{9}{9} > \left(1 - \frac{f_i(e_i)}{f_0(e_i)}\right)$) and the uniform state e_i is therefore locally asymptotically stable with respect to (4.3). From these considerations it is not at all surprising that the GA converges towards a state where all firms produce according to their second best choice. The profit of the representative firm at $t = 300$ is of course negative now, namely -0.4763.

This example can also be used to illustrate proposition 4.6.2. Assume that in a population of individuals producing $y = 0.71387$ one firm by chance decides to stop producing. As pointed out above this firm would be more successful than the others and thus we could expect that the strategy to stop producing would be imitated by about 1.33 individuals in the population. On the other hand, each individual would afterwards communicate with some other individual in the population and exchange parts of their strategies. As almost all other individuals have the strategy to produce $y = 0.71387$ this exchange will in any case lead to the abandonment of the decision not to produce. Thus, at the end of the period we expect that no more firms follow the optimal strategy to produce nothing. In other words, the number of individuals which use the optimal strategy decreases and the population again heads towards the non-equilibrium state e_i.

6.1.2 Exit and Entry Decisions

This example shows that the firms are not able to adapt to a situation where the fixed costs are too high compared to the price, because the basic decision to produce nothing at all and to avoid thereby the fixed costs can only be made by changing all bits in the string to 0. As however this basic decision changes the whole structure of the cost function, it is perhaps more plausible to decompose the decision of the firm into two parts. The firm decides first whether to produce, or not, and selects the production quantity only in the case this first decision was taken pro production.

To represent this stepwise decision making in the GA, we add a special bit to the string, whose value determines whether the corresponding firm produces, or not. Only in cases where this special bit has value 1, the quantity encoded by the other 10 bits is really produced, otherwise no production takes place. Of course this changes the situation decisively. Now we have a large number of strings encoding the decision $y = 0$, and changing from $y = y^*$ to $y = 0$ can be accomplished by changing only one single bit. The condition of proposition 4.6.2 is therefore no longer satisfied and we may anticipate that the "false equilibrium" state is no longer stable. As can be seen in figure 6.3 our theoretical considerations are confirmed by the simulations. Figure 6.3 shows the evolving behavior of the average production quantity of a GA with the extended coding in the same model as used in figure 6.2. This time however the trajectory does not approach the value $y = \frac{5}{7}$, but ends up oscillating around $y = 0.6$.

The explanation for this behavior is now straightforward. In the beginning of the run $\bar{y}$ is about 0.5, which implies that the price is about $5 - 2.5 = 2.5 < 2\sqrt{\alpha\beta} = 2$. The best decision of the firms is therefore to produce about $y = 1.25$. This induces that $\bar{y}$ increases above 0.6 and correspondingly the price falls below $p = 2$, which implies that it is now optimal to exit

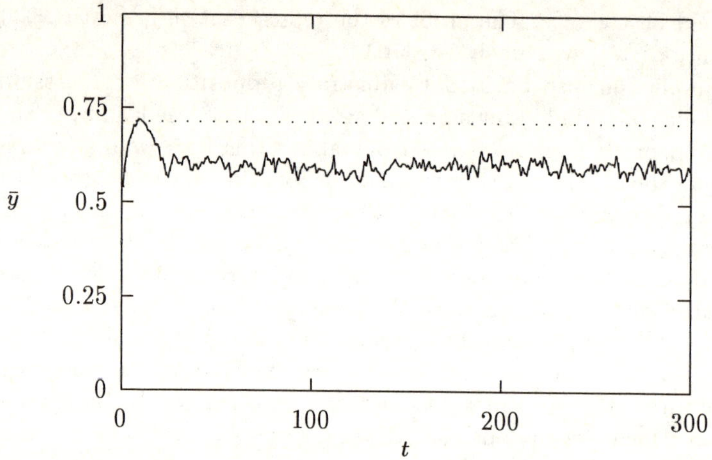

Fig. 6.3. The average production quantity in P_t for $a = 5,\ b = 5,\ \alpha = 1$ and $\beta = 1$. One special bit is added to each string saying whether the firm produces a positive quantity or nothing ($n = 1000,\ \chi = 1, \mu = 0.001$)

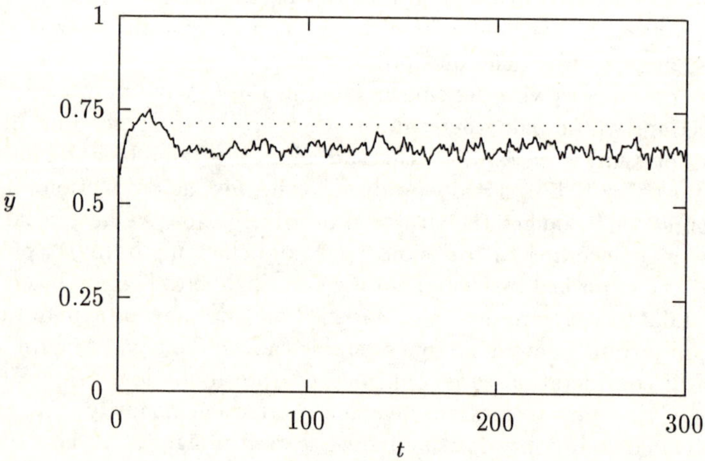

Fig. 6.4. The average production quantity in P_t for $a = 5,\ b = 5,\ \alpha = 1$ and $\beta = 1$. Two adjacent special bits describe now the decision of the firm whether to produce a positive quantity or not ($n = 1000,\ \chi = 1, \mu = 0.001$)

the market. As a shift to this optimal action needs only the changing of one single bit, several firms do this rather quickly. Accordingly the average output decreases again till the firms producing an amount larger than 0.6 have the highest fitness. Afterwards these firms are selected most and the price falls again and so on. Due to this the trajectory keeps oscillating around $y = 0.6$. Thus, we get a heterogeneous population with a rather high variance, where some of the firms produce nothing at all, while others produce a quantity near $y = \sqrt{\alpha/\beta} = 1$.

In fact the GA has found a heterogeneous rational expectations equilibrium. All firms expect the same price $p = 2\sqrt{\alpha\beta}$, but for this expected price the optimal production quantity is according to (6.3) not unique. Both $y = 0$ and $y = \sqrt{\alpha/\beta}$ yield the optimal profit of 0. If the fraction of firms producing $y = \sqrt{\alpha/\beta}$ equals exactly $x = \frac{a}{b}\sqrt{\beta/\alpha} - 2\frac{\beta}{b}$ (it is easy to show that $x \in [0,1]$ if (6.2) holds and (6.5) does not hold) we get from (6.1)

$$p = a - b\left(\frac{a}{b}\sqrt{\frac{\beta}{\alpha}} - 2\frac{\beta}{b}\right)\sqrt{\frac{\alpha}{\beta}} = a - a + 2\sqrt{\alpha\beta} = 2\sqrt{\alpha\beta}.$$

Therefore all firms did expect the right price. We have a heterogeneous rational expectations equilibrium where all firms hold the same expectations about the price, which hold true if the firms take different optimal actions in a given proportion (formally speaking a state is called a heterogeneous rational expectations equilibrium if $p_i^e = p^* \ \forall i = 1 \dots n, y_i \in y^*(p^*) \ \forall i = 1 \dots n$ and $p^* = a - \frac{b}{n}\sum_{i=1}^{n} y_i$ holds). This heterogeneous rational expectations equilibrium emerges from the homogeneous equilibrium as the fixed costs pass the value of $\alpha = \frac{\beta a^2}{(2\beta+b)^2}$. Beyond this value, a rational expectations equilibrium can only be adopted if some firms retire from the market and do not produce any more. If enough firms exit from the market the price of the good is sufficiently high such that the firms producing the optimal quantity gain a profit of zero and a heterogeneous rational expectations equilibrium is reached. The number of firms producing in the equilibrium decreases for increasing fixed costs until for $\alpha = \frac{a^2}{4\beta}$ no firm is in the market anymore.

For our parameter values we get $x = 0.6$, which means that 400 of 1000 firms should decide not to produce. In the simulation with $n = 1000$ depicted here, we observe that after 300 generations 378 out of the 1000 firms have stopped producing. The average profit in the population should be 0, the actual average profit in the simulation shown in figure 6.3 is 0.0262. Considering the disruptions due to mutations and crossover, the approximation of the heterogeneous equilibrium state is quite satisfactory.

Recalling proposition 4.2.1, the learning of a heterogeneous equilibrium seems to be a little strange as we have shown that the trajectory of the GA is close to uniform states most of the time. Note however that a population corresponding to a heterogeneous rational expectations equilibrium is almost uniform. The strings representing producing firms differ in general only in the first bit value from strings representing a firm which has exited the market.

Thus, a population corresponding to a heterogeneous rational expectations equilibrium is uniform in 10 of the 11 bit positions and has only some variance in one single position. Therefore, there is no contradiction with proposition 4.2.1.

To further illustrate the relevance of the theoretical results, I present now two additional simulations with slightly modified coding mechanisms. In figure 6.4 another special bit is added to the binary string. This second special bit is located adjacent to the first special bit at the end of the string. Now these two bits represent the decision whether to produce or not to produce. If both bits have value 0 the firm does not produce in the current period. In any other case the firm produces, and both bits are set to value 1 before the selection at time $t+1$ is made. On the other hand, the mutation probability for these two bits is $\sqrt{\mu}$ instead of μ. This means that the probability that a firm which produces in the current period will — due to mutations — exit the market in the next period is the same as in the last simulation (the change needs now two mutations, but the probability that these occur at the same time is still μ). The length between the two bits which distinguish the strings encoding $y = y^*$ from the strings encoding $y = 0$ is 1. This implies by proposition 4.6.2 the instability of the uniform state corresponding to the rational expectations equilibrium for lower fixed costs. We observe in figure 6.4 that the change of coding does not change the behavior of the GA very much. The oscillations around the heterogeneous equilibrium are a little larger, since the probability for the change from not producing to producing is larger in this setup. Nevertheless we get qualitatively the same result as in figure 6.3, namely the instability of the homogeneous "local" equilibrium.

To see how important the position of the bits which distinguish the string with the highest fitness from the strings predominant in the population is for the behavior of the GA, we have to change the coding once again. Two special bits are used again for the encoding of the basic production decision, but this time the two bits are situated at the opposite ends of the string. Let $i \in \Omega$ be a string encoding $y = y^*$ and j be a string encoding $y = 0$, then the length between the two outmost differing bits is $l - 1$. Therefore, we may anticipate from proposition 4.6.2 that the uniform state e_i is locally stable if this coding is used.

In figure 6.5 we can see the trajectory of the average production quantities in P_t for the parameter values (6.7). As in figure 6.2 the GA settles into a state where all firms act according to the second best reply to the market price. The state is obviously stable, for the continuous disruptions of the homogeneous state caused by mutations never lead the population state away from e_i. As we have seen in figure 6.4, mutations with the same frequency may however lead to a long-lasting deviation from e_i, if the two special bits are adjacent. We observe very clearly the importance of the coding mechanism in this model. In the case where no uniform rational expectations equilibrium exists, the coding of the decision of the firm determines whether the population will

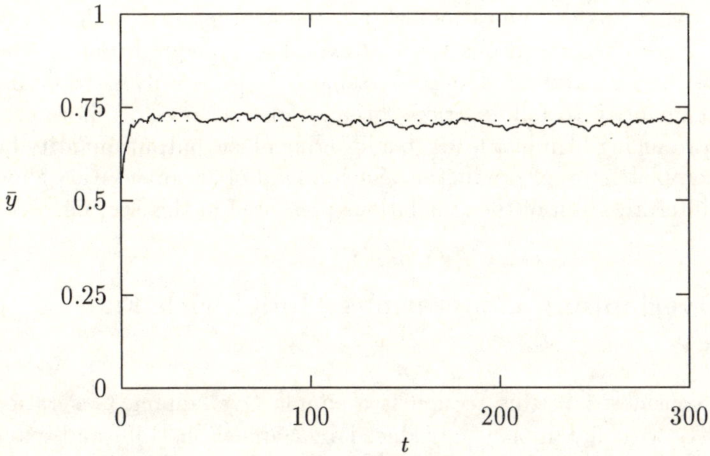

Fig. 6.5. The average production quantity in P_t for $a = 5,\ b = 5,\ \alpha = 1$ and $\beta = 1$. Two special bits at the opposed ends of the string describe now the decision of the firm whether to produce a positive quantity or not ($n = 1000,\ \chi = 1, \mu = 0.001$)

settle into a "local" equilibrium, where all firms act according to the second best choice and make a negative profit, or at a heterogeneous state where all anticipate the right price and by acting optimal make a profit of zero.

The condition that the two special bits — which are in some sense the most important bits — have to be close to each other in order to facilitate the learning of the heterogeneous rational expectations equilibrium reminds a little of the schema theorem and the building block hypotheses. Here, however, the importance of the special bits for the stability of e_i lies in the fact that they distinguish the best from the second best choice in the state e_i. The only correct way to interpret the results of the simulations shown in figure 6.1–6.5, is therefore the reference to the results of chapter 4. As already mentioned before the idea of a building block makes in general no sense in economic systems, since the fitness of a schema depends on the state of the population.

The dependence of the limit behavior of the GA from the chosen coding of actions underlines one of the problems of the use of GAs in economic systems. In order to decide which of the two simulations, the one of figure 6.4 or the one of figure 6.5, shows the more reasonable behavior, we have to interpret the differences in the setup of the two simulations from an economic point of view. This does not seem to be possible, but the coding of a single yes or no decision with two bits is itself rather questionable and was mainly done to illustrate the theoretical results. If we compare the setups used in the simulations of figures 6.2 and 6.3, we are able to interpret these results also in an economic sense. If the firms treat the basic question of production or non-production separately, they may adapt, due to information from non-producing other firms or innovations within the firm, quickly to low prices and end up in the heterogeneous rational expectations equilibrium. If they do not make the exit and entry decision separately, they only slowly adapt their output quantity to falling prices, because an exchange of information with a non-producing firm just leads to a lowering of the output quantity but never to a complete stop of production. Such a kind of argumentation allows a plausible interpretation of the simulations presented in this section.

6.2 An Overlapping Generations Model with Fiat Money

The model considered in this section is a simple Overlapping Generations Model (OLG) with fiat money introduced by Samuelson [148] and extensively analyzed afterwards[2]. Azariadis [17] and other authors have shown that in OLG models rational expectations equilibria with random prices may exist although there is no uncertainty within the model. These equilibria, which are governed by the beliefs of the individuals that an external random variable has some influence on the prices, are called sunspot equilibria. Woodford [186] showed that under certain conditions such equilibria are learned with probability one by a stochastic approximation learning rule. On the other hand Laffond and Lesourne [105] have introduced a simple adaptation process which does not necessarily lead to convergence against the sunspot equilibrium. Arifovic [9] showed that in models without cyclical or sunspot equilibria a GA always converges towards the monetary steady state.

It will become clear when the model is introduced that two binary populations are needed to model the interaction between the different generations. Nevertheless, this is no two population model in the sense of section 4.7. The reason for this will be made clear later on. Due to the complicated structure of this model no analytical explanation for the simulation results can be given in this section. Instead, numerical experiments will be used to address the following questions:

(i) Which equilibrium is learned by a GA in models where the two stationary steady states and also periodic equilibria with period two exist?

(ii) Is a GA able to learn sunspot equilibria if the coding of the actions induces a dependence of the actions of the individuals from the sunspot variable?

Consider a population where each individual lives for exactly two periods. Time evolves in discrete steps from 0 to ∞ and at any time t there are n young

[2] The results presented in this section were first published in Dawid [46].

and n old individuals. Each individual is endowed with e_1 units of the only existing consumer-good in youth and with e_2 in old age. As the consumer-good is not durable, the only possibility to transfer wealth form youth to old age is by the means of fiat money. Fiat money is intrinsically worthless and only used for trading between the generations. To establish the fiat money in the economy, we assume that at time 0 every old individual is not only endowed with e_2 units of the consumer-good, but also with one unit of fiat money. All individuals have the same preferences, which are represented by a utility function $U(c_1, c_2)$, where c_1 is the consumption in youth and c_2 the consumption in old age. Assume that the utility function has the form $U(c_1, c_2) = u_1(c_1) + u_2(c_2)$ and that u_1 and u_2 are concave increasing. The assumption that no cross dependencies between utility in youth and old age exist is often made in the literature (see Woodford [186], Lucas [111]). Let $y_{i,t}$ denote the supply of a young individual i at time t. As the representative old individual owns one unit of money the equilibrium condition on the goods market is written as

$$\frac{n}{p_t} = \sum_{i=1}^{n} y_{i,t},$$

where p_t is the price of the consumer-good at time t. It is assumed that the young can not borrow (i.e. $y_{i,t} \geq 0$) and accordingly we have $p_t = \frac{1}{\bar{y}_t}$, where $\bar{y}_t$ is the average supply of the young individuals at time t. Now let $r_t = \frac{p_t}{p_{t+1}}$ be the real return of fiat money. The young individual does neither know p_t nor p_{t+1} which implies that at time t the variable r_t is not certain for the young individuals. Nevertheless they have an expectation about r_t and this expectation is denoted by $\rho_{i,t}$. The young individual i expects to get $\rho_{i,t}$ units of the consumer-good in old age for one sold in youth. Writing the utility of an individual as

$$V(r, y) := U(e_1 - y, e_2 + ry),$$

we get that a young utility maximizing individual, who expects a real return of $\rho_{i,t}$, chooses his supply according to the following rule:

$$y_{i,t} = s(\rho_{i,t}) := \text{argmax}_{0 \leq y \leq e_1} V(\rho_{i,t}, y).$$

The function $s(r)$ is called the savings function of the representative individual. The elasticity of the savings with respect to r will be denoted by $\epsilon(r)$.

A (homogeneous) competitive equilibrium with perfect foresight is given by a sequence of prices $\{p_t\}_0^\infty$ satisfying $s\left(\frac{p_t}{p_{t+1}}\right) = \frac{1}{p_t}$. A stationary equilibrium with perfect foresight or a (homogeneous) rational expectations equilibrium is a situation where all young individuals have perfect foresight and always supply the same optimal amount. Thus, the equilibrium price is given by the solution of the equation

$$p\, s(1) - 1 = 0. \tag{6.8}$$

There are at most two rational expectations equilibria in the model, the monetary steady state given by $p_t \equiv \frac{1}{s(1)}$ and in cases where $\lim_{c \to 0} cu_2'(c) = 0$ also the non-monetary steady state with $\bar{y}_t \equiv 0$, where no trading happens at all. It has been shown, that the introduction of an adaptive learning scheme in order to build the expectation about $\rho_{i,t}$ does in several cases lead to convergence to the monetary steady state (Woodford [186], Benassy and Blad [20], Lucas [111], Arifovic [9]). A (homogeneous) periodic competitive equilibrium with perfect foresight of order 2 is given by a sequence of prices with $p_t = p_{t+2} \neq p_{t+1}$ $\forall t$, which satisfy

$$\frac{1}{p_t} - s\left(\frac{p_t}{p_{t+1}}\right) = 0$$
$$\frac{1}{p_{t+1}} - s\left(\frac{p_{t+1}}{p_t}\right) = 0.$$

In order to find a periodic equilibrium of order 2 we have to solve the following equation:

$$rs(r) - s\left(\frac{1}{r}\right) = 0. \tag{6.9}$$

Now let us introduce sunspots to our system, i.e. we assume that the individuals believe that the prices in the system and also the real return of money depends on an exogenous random variable. This random variable, which will be denoted by τ has no intrinsic influence on the system parameters like the endowments e_1 and e_2 or the utility function. Nevertheless, the beliefs of the individuals may be correct if the whole population believes that the sunspot variable will influence the system and acts according to this belief. The sunspot variable, follows a Markov process on the state space $\{0, 1\}$ defined by the transition matrix

$$\Pi = \begin{pmatrix} \pi_{00} & \pi_{01} \\ \pi_{10} & \pi_{11} \end{pmatrix}.$$

The value π_{ij} denotes the probability for $\tau_{t+1} = j$ under the condition that $\tau_t = i, \quad i, j \in \{0, 1\}$. Of course Π is a stochastic matrix which means that $\pi_{i0} + \pi_{i1} = 1, \ i = 0, 1$. All individuals believe that the prices depend only on the value of τ, which means that an individual i anticipates a price $p0_i^e$ whenever $\tau = 0$ occurs and $p1_i^e$ when the sunspot variable equals 1. As the value of τ_{t+1} is not known at time t a young individual has to maximize his expected utility under $\tau_t = 0, 1$, which yields

$$y_{i,t} = \begin{cases} z\left(\frac{p0_i^e}{p1_i^e}, \pi_{00}\right) & \tau_t = 0 \\ z\left(\frac{p1_i^e}{p0_i^e}, \pi_{11}\right) & \tau_t = 1 \end{cases}, \tag{6.10}$$

where

$$z(r, \pi) := \arg \max_{0 \leq y \leq e_1} \left[\pi V(1, y) + (1 - \pi) V(r, y)\right]$$

is the counterpart of the savings function for sunspot believers. We say that the pair $(p0, p1)$ defines a stationary sunspot equilibrium (SSE) with respect to the transition matrix Π, if all individuals believe in sunspots and the expected prices coincide with the actual realizations of the prices $p0$ and $p1$. This means that we have to find a solution ω^* of

$$\omega z(\omega, \pi_{00}) - z\left(\frac{1}{\omega}, \pi_{11}\right) = 0 \tag{6.11}$$

and set

$$p0 = \frac{1}{z(w^*, \pi_{00})} \qquad p1 = \frac{1}{z\left(\frac{1}{w^*}, \pi_{11}\right)}.$$

It is obvious that a period 2 equilibrium can be interpreted as a SSE with respect to a transition matrix Π given by $\pi_{ii} = 0,\ \pi_{ij} = 1, i \neq j$. Azariadis and Guesnerie [18] showed that the connection between cyclical equilibria and sunspot equilibria is very close. Actually they proved that a two state stationary sunspot equilibrium exists if and only if a periodic competitive equilibrium with perfect foresight of order 2 exists. Further they derived the following sufficient condition for the existence of sunspot equilibria:

$$\epsilon(1) < 0 \text{ and } \pi_{00} + \pi_{11} < 2 - \frac{1}{|\epsilon(1)|}. \tag{6.12}$$

Application of this result to cyclical equilibria of period 2 (i.e. $\pi_{00} = \pi_{11} = 0$) yields that

$$\epsilon(1) < -\frac{1}{2}$$

is sufficient for the existence of periodic competitive equilibria of order 2.

In our simulations each binary string P_t encodes the expectation of one young individual about the real return $\rho_{i,t}$. We denote by $\rho(k)$ the expectation encoded by the string $k \in \Omega$ which means that $\rho_{i,t} = \rho(k)$ if individual i is represented by the binary string $k \in \Omega$ in P_t. We assume that all individuals are able to determine the optimal quantity to supply given their expectation about the real return (i.e. $y_{i,t} = s(\rho_{i,t})$). Therefore the price at time t is given by

$$p_t = \frac{n}{\sum_{k \in P_t} s(\rho(k))}.$$

In order to calculate the utility gained by an individual expecting $\rho(k)$, we have to know r_t and accordingly p_{t+1}. For this reason the population of young individuals at time $t+1$ has to be represented by a separate second population. The state of the second population determines p_{t+1} and thus also r_t. The utility of an individual is now given by $V(r_t, s(\rho(k)))$, and it is easy to see that this value does not only depend on the state of P_t but also on the state of P_{t+1}. This implies that the state of the population at time $t+2$ depends on the states at time t and $t+1$. However, the state at time $t+3$

depends on the states at time $t+1$ and $t+2$. Thus the dynamics for large populations is given by

$$(\phi^{t+2}, \phi^{t+3}) = (\mathcal{M}^1(\mathcal{S}^1(\phi^t, \phi^{t+1})), \mathcal{M}^2(\mathcal{S}^2(\mathcal{M}^1(\mathcal{S}^1(\phi^t, \phi^{t+1})), \phi^{t+1}))).$$

Obviously the structure of this dynamical system is quite different from the one in (4.8) and thus the results of section 4.7 are not valid in this setup.

The fitness value of the string k is determined by applying a linear scaling mechanism as described in chapter 3. The parameter c governing the selection pressure has in our simulations always the value $c = 1.5$. In addition to the three standard operators I use in the simulations in this model also the election operator described in section 3.3.4. The election operator reduces the variance in the population and leads to a smoother approach towards the equilibrium. If for example the parent strings encode a price expectation lower than the actual price, election will accept only an offspring which lies in between the parents' expectation and the real price, but it will prevent a string with an even smaller price expectation from entering the population. This kind of control operator has proved to facilitate the learning of equilibria in this model, but the results are qualitatively the same without election.

As already mentioned above, this setup does not allow a theoretical analysis of the behavior of the GA any more, and we have to rely entirely on simulations. As known from the simulations of Arifovic [10] the GA has no problems in learning the steady state equilibrium if no cyclical equilibria exist. Therefore, here we deal with a model where an equilibrium with period two and therefore also a sunspot equilibrium exists.

6.2.1 Learning of Cyclical Equilibria

We know that under standard assumptions the condition $\epsilon(1) < -\frac{1}{2}$ is sufficient for the existence of cyclical equilibria with period two. Obviously, if $s(r)$ equals neither 0 nor e_1, we have the following equality for all values r:

$$V_2(r, s(r)) = 0.$$

Implicit differentiation yields

$$s'(r) = -\frac{V_{12}(r, s(r))}{V_{22}(r, s(r))},$$

and therefore

$$\epsilon(1) = \frac{s'(1)}{s(1)} = -\frac{u_2''(e_2 + s(1))}{u_2''(e_2 + s(1)) + u_1''(e_1 - s(1))}$$
$$-\frac{u_2'(e_2 + s(1))}{s(1)(u_2''(e_2 + s(1)) + u_1''(e_1 - s(1)))}.$$

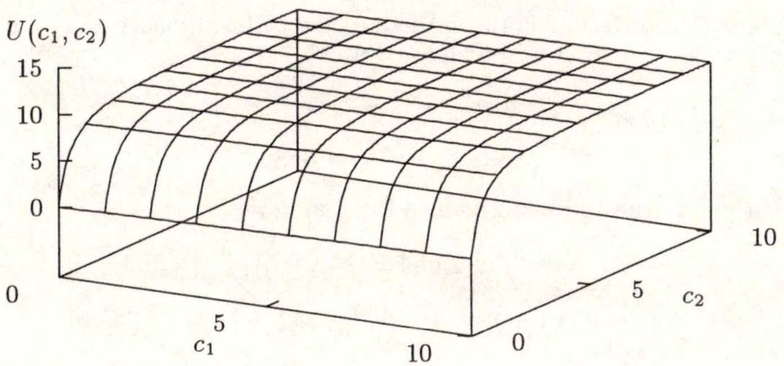

Fig. 6.6. The utility function $U(c_1, c_2)$

It follows from this expression that the sufficient condition for cycles can never be met when $u_1(c) = u_2(c)$. The first of the two summands is negative the second positive, and in order to obtain cyclical equilibria we thus have to choose u_1 and u_2 in a way that the first summand has a relatively large absolute value compared to the second. This implies that the concavity of u_2 has to be large compared to the concavity of u_1. The utility function

$$U(c_1, c_2) = 0.1c_1^{0.9} + 10 - \frac{10}{(1 + c_2)^2}. \tag{6.13}$$

meets these requirements and is used in all simulation runs presented in this subsection. The graph of the function $U(c_1, c_2)$ is shown in figure 6.6. It can be clearly seen in the figure that the concavity with respect to c_1 is much smaller than the concavity with respect to c_2. The parameter values used are $e_1 = 10$ and $e_2 = 0$, which yields $\epsilon(1) = -0.578$. The monetary steady state is given by

$$p^* = \frac{1}{s(1)} = 0.186,$$

which implies

$$y^* = s(1) = 5.374 \qquad V(1, y^*) = 10.150.$$

The real return of money in the competitive equilibrium with period 2 is given by

$\bar{p}_{399}$:	4.780	$\bar{p}_{400}$:	0.170
$\bar{y}_{399}$:	2.075	$\bar{y}_{400}$:	9.975
$\bar{f}_{399}$:	10.561	$\bar{f}_{400}$:	8.946
p_{399} :	0.482	p_{400} :	0.100
r_{399} :	4.807	r_{398} :	0.208

Table 6.1. The results of the GA in a model with stationary and cyclical equilibria with $n = 30,\ l = 20,\ \chi = 0.6$ and $\mu = 0.1$

$$r^* = 4.868,$$

which yields the following values for y and V:

$$s(r^*) = 2.054\ V(r^*, s(r^*)) = 10.563$$
$$s\left(\tfrac{1}{r^*}\right) = 10 \quad V\left(\tfrac{1}{r^*}, s\left(\tfrac{1}{r^*}\right)\right) = 8.928.$$

It is observed that one of the two values for y in the cycle equals the highest possible value e_1. This makes it more difficult for the GA to learn the equilibrium, for any individual which anticipates a wrong real return slightly larger or smaller than r^* also sells his complete endowment e_1, which means that these non-equilibrium strings have the same fitness as the strings representing individuals with beliefs which are correct in the equilibrium.

The numerical experiments were run for 400 generations with a relatively small population size ($n = 30$), and a relatively small crossover probability ($\chi = 0.6$), which means that the exchange of information is done only reluctantly in the population. As can be seen in figure 6.7, already after 100 generations the GA has converged quite accurately to the equilibrium with period 2. Several simulations were also made for different values of μ between $\mu = 0.01$ and $\mu = 0.4$ and also for different values of the scaling parameter c. However, no qualitative changes in the results were observed and I abstain from presenting all these results here. It is apparent from the simulations that the monetary steady state is unstable for the dynamics induced by the GA, whereas the cyclical equilibrium is attracting.

One of the two populations adopts a very small expectation about the real return of money and sells all its consumer-goods in youth. This behavior is exploited by the old generation, which saves only little money under the correct assumption that the goods will be rather cheap in their old age. It can be seen rather easily why such a pattern evolves. Note that the concavity of the utility in youth is very weak, which means that the marginal disutility of sold consumer-goods in youth is almost constant. As in the optimum the marginal disutility in youth must equal the marginal utility in old age, we realize that the savings function is in our case approximately given by $ru_2'(rs(r)) = \text{const}$. It can be easily seen that the function $ru_2'(ry)$ is always nonincreasing in y, and also nonincreasing in r if $1 - 2ry < 0$. This inequality however is fulfilled with a large probability in the beginning of the run. As the initialization is done randomly, we may anticipate a real return of about

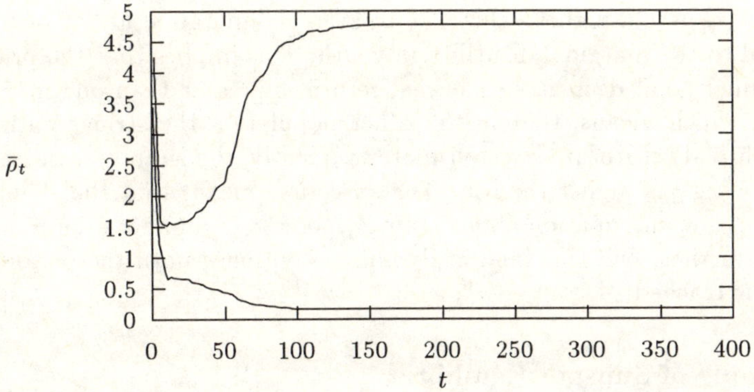

Fig. 6.7. The evolution of the population averages of the expected real return of money in both populations ($n = 30$, $l = 20$, $\chi = 0.6$ and $\mu = 0.001$)

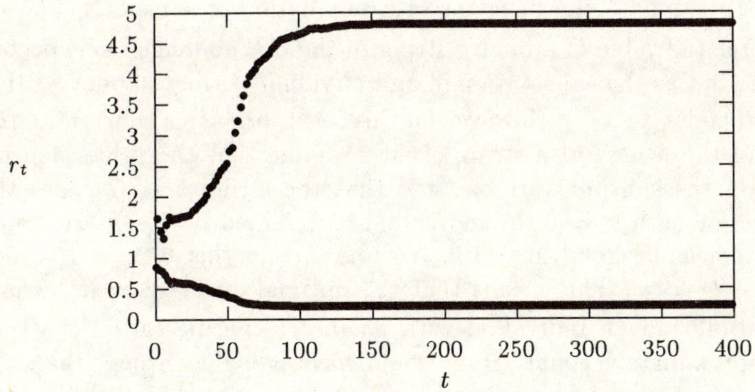

Fig. 6.8. The observed real return of money for the same simulation as in figure 6.7

1 (see also figure 6.8). If in this situation — due to stochastic effects — in one population P_t most of the individuals anticipate a real return smaller than one, they select a higher value of y to keep the marginal utility of the old age equal to the marginal disutility in youth. This implies that the price of the consumer-good decreases. The real return is greater than one in the next period, which means, that in the other population the strings anticipating a high real return are selected more frequently, the supply decreases, and the price p_{t+1} is higher than p_t. This of course means that the strings representing a low anticipation about r_t in P_t increase in number even more, p_t decreases further, and this kind of dynamics continues until the periodic equilibrium is reached.

6.2.2 Learning of Sunspot Equilibria

Having seen that the existence of a periodic equilibrium implies that a GA rather approaches this cycle than the stationary equilibrium let us now deal with the question whether also sunspot equilibria may be learned. As the existence of equilibria with period 2 is sufficient for the existence of sunspot equilibria, we may again use the utility function given by (6.13) and $e_1 = 10, e_2 = 0$. Condition (6.12) yields that a sunspot equilibrium exists with respect to any transition matrix Π with $\pi_{11} + \pi_{22} < 0.27$. In order to learn sunspot equilibria we have to assume that all individuals believe that the prices depend on the value of the sunspot variable. From (6.10) follows that, given Π, the action of such a "sunspot believer" i depends only on the value $\omega_{i,t} = \frac{p0_i^e}{p1_i^e}$. To simulate the behavior of a population of sunspot believers, let us interpret the value of a binary string in the GA no longer as expected value of $\frac{p_{t+1}}{p_t}$, but as the expectation of an individual at time t about $\frac{p0}{p1}$ (i.e. each string encodes an $\omega_{i,t}$). However, in this setup problems concerning the calculation of the fitness of a string occur. Assume that the prices depend completely on the sunspot variable, and that the string $k \in \Omega$ codes the correct value for w. If $\tau_t = 0$ the individual represented by k sells $z(w, \pi_{00})$ units of the consumer-good in youth. Assume further that $\tau_{t+1} = 1$, which means that $\frac{p_t}{p_{t+1}} = w$. This means that the optimal quantity to sell would have been $s(w)$, and an individual with an incorrect expectation $\tilde{\omega}$ where $z(\tilde{\omega}, \pi_{00})$ approximately equals $s(w)$ would have a higher fitness than the individual with the correct expectation. To circumvent this problem, the genetic operators are applied only every 40 generations and the fitness of a string is defined as the average utility over these 40 generations. As this value approximates the expected utility of an individual, the strings coding good estimations of the observed values of w have the highest fitness values.

In the simulations the transition matrix

$$\Pi = \begin{pmatrix} 0.1 & 0.9 \\ 0.9 & 0.1 \end{pmatrix}$$

τ_{2999} :	1	τ_{3000} :	0
$\bar{\omega}_{2999}$:	2.317	$\bar{\omega}_{3000}$:	2.155
$\bar{y}_{2999}$:	8.005	$\bar{y}_{3000}$:	3.664
$\bar{f}_{2999}$:	10.148	$\bar{f}_{3000}$:	10.403
p_{2999} :	0.124	p_{3000} :	0.273
r_{2999} :	0.458	r_{2998} :	2.185

Table 6.2. The results of the GA incorporating sunspot depending expectations with $n = 30$, $l = 20$, $\chi = 0.6$ and $\mu = 0.1$.

was used. The SSE with respect to this transition matrix is given by

$$w^* = 2.37 \qquad p0^* = 0.285 \qquad p1^* = 0.12$$

$$z\left(\frac{p0^*}{p1^*}, 0.1\right) = 8.321 \qquad z\left(\frac{p1^*}{p0^*}, 0.1\right) = 3.504.$$

This time 3000 generations were calculated, which means that the genetic operators were applied 75 times. Figures 6.9 and 6.10 show the evolution of the expectations $\bar{\omega}_t$ and the observed real returns of money.

From figure 6.10 it becomes obvious that after a short transient period the prices really depend only on the value of the sunspot variable. Note that the values near 1 indicate periods where the sunspot variable attained the same value for two consecutive periods. This shows that we have a true sunspot equilibrium where the observed values of the prices are completely governed by the sunspot variable. This must be especially emphasized, for it could be assumed that due to the fact that two populations have to be used in our model a cycle of period two may be the natural outcome of such a simulation. Figure 6.10 shows that this is not the case. The results after these 3000 generations, given in table 2, show that the values obtained in the simulations deviate a little more from the equilibrium values than in the case of the periodic equilibrium. The average deviation of the observed value of $\frac{p0}{p1}$ from the equilibrium value w^* is 0.28 or 12% compared to less than 6% observed in the simulations with a periodic equilibrium (of course we had to compare $\frac{p_t}{p_{t+1}}$ with $\hat{r}$ in the periodic case). However, this larger deviation had to be expected as the fitness function is now stochastic.

Summarizing we may say, that this second example of genetic learning in an economic system shows the ability of GAs to find also non steady state equilibria in systems with a more complicated dynamics than the one described in chapter 4. Maybe future work will enable us to extend the theoretical results of chapter 4 to systems with an overlapping generations structure and explain the obtained results also from an analytical point of view. In the next section we will see how the existing results may be applied in multi-population models without an overlapping generations structure.

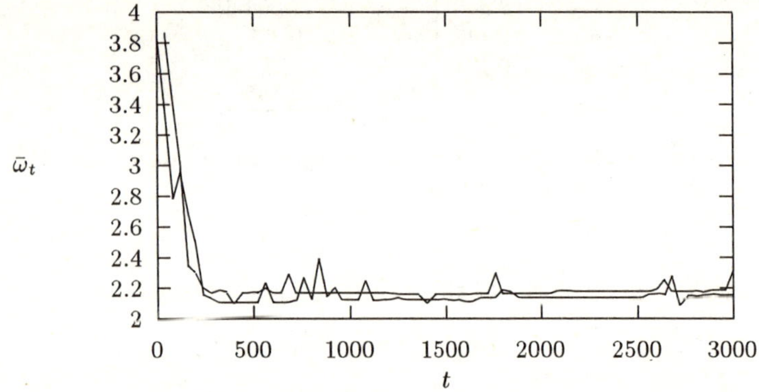

Fig. 6.9. The evolution of the population averages of the expected value of $\frac{p0}{p1}$, when both populations believe in sunspots ($n = 30$, $l = 20$, $\chi = 0.6$ and $\mu = 0.001$)

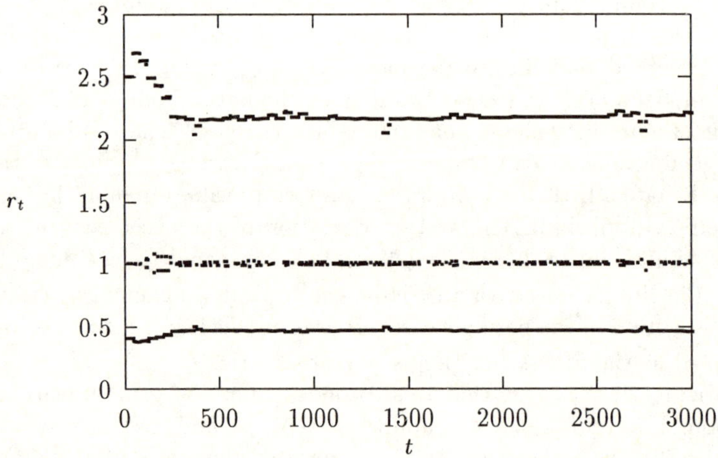

Fig. 6.10. The observed real return of money for the same simulation as in figure 6.9

6.3 A Sealed Bid Double Auction Market

Double auction models are of major importance for the understanding of price formation in markets and sealed bid double auctions have been analyzed extensively in the literature from a theoretical (e.g., Chaterjee and Samuelson [38]) as well as from an empirical (e.g., Radner and Schotter [138]) point of view. In the setup considered here there are two populations of agents called buyers and sellers. Both populations contain the same number of agents and this number is denoted by $n > 0$. A single good is traded in the market. Each seller has the potential to produce one unit of this good every period. The production costs are given by $c_j^t \in [0,1]$ for seller j in period t and the seller produces the good only if he can sell it in the same period. On the other hand, buyer i gains utility of $u_i^t \in [0,1]$ from consuming the good. The values of u_i^t and c_j^t are private information. During each period, every seller is randomly matched with a buyer and both submit a sealed bid. Depending on whether the values of u_i^t and c_j^t are stochastically determined every period or stay constant throughout all periods the agents use different types of bidding functions. In the first case the agents' bidding behavior is determined by a map from $[0,1] \to [0,1]$ which gives the bid respectively ask in dependence on the current value of u_i^t respectively c_j^t. We will deal with this case later on and assume now that all utility values are the same, all cost values are the same and that these two values are randomly chosen at the beginning of period one and never changed afterwards.

6.3.1 Constant Cost and Utility Values

Here buyers and sellers are homogeneous in the sense that all buyers have the same utility value and all sellers have the same piece costs[3]. These values also do not change from period to period. This means that there are two real values u and c with $u_i^t = u \;\forall t,\; i = 1,\ldots,n, \quad c_j^t = c \;\forall t,\; j = 1,\ldots,n$. In order to allow any trade on this market it has to be assumed that $u > c$. In such a situation the bidding behavior of the agents can be simply represented by the actual bid respectively ask. The functional dependence on c and u can be neglected, since these values are now fixed parameters of the model which are the same for all agents. The buyer submits the price he is willing to pay (p^b), and the seller gives the minimum payment for which he will deliver the good (p^s). Buyers and sellers know that c and u lie in $[0,1]$ and accordingly restrict their bids to this interval. If $p^b \geq p^s$ one unit of the good is traded at a price of

$$p^{tr} = \frac{p^b + p^s}{2},$$

if $p^b < p^s$ no trade takes place. Accordingly, the payoffs of buyers and sellers read

[3] This model has been analyzed in Dawid [48].

$$\Pi^b(p^b,p^s) = \begin{cases} u - \frac{p^b+p^s}{2} & p^b \geq p^s \\ 0 & p^b < p^s \end{cases}$$

$$\Pi^s(p^b,p^s) = \begin{cases} \frac{p^b+p^s}{2} - c & p^b \geq p^s \\ 0 & p^b < p^s \end{cases}$$

Since agents do not know with whom they will be matched with in the current period, their payoffs a priori are stochastic. Of course the expected utilities depend on the population distribution in the other population. Let $P^s = \{p_1^s, \ldots p_{m^s}^s\}$, $p_i^s \in [0,1], p_i^s < p_{i+1}^s$ be a finite set of sellers' asks and $s^s = (s_1^s, \ldots, s_{m^s}^s) = \frac{1}{n}(x_1^s, \ldots, x_{m^s}^s)$, $x_i^s > 0, x_i^s \in \mathbb{N}_0, \sum_{i=1}^{m^s} x_i^s = n$ a population distribution over P^s. Then, the expected utility of a buyer setting price $p^b \in [0,1]$ is given by

$$U^b(p^b, P^s, s^s) = \sum_{i:p_i^s \leq p^b} s_i^s \Pi^b(p^b, p_i^s). \tag{6.14}$$

On the other hand, using analogous definitions for the buyers' offers

$$U^s(p^s, P^b, s^b) = \sum_{i:p_i^b \geq p^s} s_i^b \Pi^s(p_i^b, p^s) \tag{6.15}$$

is the expected utility of a seller with asking price p^s if the distribution of the buyers' offers is given by s^b over P^b. The system described above is in equilibrium if any buyer chooses a price which maximizes the expected payoff against the sellers' population distribution, and every seller chooses a price which maximizes the expected payoff against the buyers' population distribution. Stated more formally:

Definition 6.3.1. *Let P^b and P^s be finite sets of buyers' offers and sellers' asks. Further, let s^b be a population distribution over P^b, and s^s be a population distribution over P^s. The quadruple (P^b, P^s, s^b, s^s) is an equilibrium of the double auction market if*

$$U^b(p^b, P^s, s^s) \geq U^b(p, P^s, s^s) \quad \forall p^b \in P^b,\ \forall p \in [0,1]$$

and

$$U^s(p^s, P^b, s^b) \geq U^s(p, P^b, s^b) \quad \forall p^s \in P^s,\ \forall p \in [0,1].$$

Given this definition, we get the following characterization of the equilibria of the (static) double auction market (a proof of the proposition is given in Appendix C).

Proposition 6.3.1. *Let P^b and P^s be finite sets of buyers' offers and sellers' asks. If there exist population distributions s^b and s^s such that (P^b, P^s, s^b, s^s) is an equilibrium then $P^b = P^s = P$, $m^b = m^s = m$ and $c \leq p_1, u \geq p_m$. On the other hand, if $P^b = P^s = P$, where P has only rational elements with $c \leq p_1, u \geq p_m$ and u and c are rational then there exists a population size n and population distributions s^b and s^s, such that (P^b, P^s, s^s, s^b) is an equilibrium.*

Note that proposition 6.3.1 does not state which equilibria exist for a given population size n. Most probably the number of equilibria varies substantially for different values of n. The only equilibria which obviously exist for all population sizes are those where $P = \{p\}$. In these cases we can also abandon the condition that p is a rational number and get

Corollary 6.3.1. *If* $P^b = P^s = \{p\}$, $p \in [c, u]$ *then* $(P^b, P^s, 1, 1)$ *is an equilibrium of the double auction market.*

This corollary is particularly interesting, if we think of proposition 4.2.1 where it has been shown that genetic learning in the long run stays close to uniform states most of the time. This model has a huge number of equilibria and also a large number of uniform equilibria, so we might expect that the problem of equilibrium selection is rather hard. But before I can address this problem I have to explain how the genetic algorithm learning works in this setup.

Each of the populations the buyers and the sellers is represented by a population of binary strings. All strings have the same length l and each binary string represents one agent. Since the population size and string lengths in both populations coincide we may neglect the indices indicating the population number when considering the set of binary strings in a population and the set of population states. Thus, I will write $\Omega := \Omega^1 = \Omega^2$, $r := r^1 = r^2$ and $S := S^1 = S^2$. An agent is completely described by his bidding price, which is encoded by the corresponding binary string. The offer respectively ask encoded by $j \in \Omega$ is given by

$$p(j) = \sum_{i=1}^{l} j(i) \frac{1}{2^i}.$$

Every period each string in the population of buyers is randomly matched with a string in the population of sellers. If the buyer is represented by string $j \in \Omega$ and the seller by $k \in \Omega$ the buyer receives a payoff $\Pi^b(p(j), p(k))$ and the seller $\Pi^s(p(j), p(k))$. Both populations are initialized randomly (in a few cases specially chosen deterministic initializations are used; see below) and the matching procedure described above is repeated for T periods, where all stochastic events in different periods are stochastically independent. The bids and asks are not determined anew every period but only every m periods the agents choose a new bidding price. The selection of a new bidding price is modelled by the application of the three standard genetic operators proportional selection, one-point crossover and mutation to both populations of strings. The fitness of a string is simply the average payoff it earned in the last m periods. Note that a priori the case where offers are higher than u or lower than c are not excluded. However, strings yielding a negative average payoff get assigned a fitness of zero and thus should quickly become extinct. Due to the random matching procedure the fitness of a string is a random variable where the expected value depends on the value of the string and the current population state of the opponents' population. Of course, the larger m is the smaller is the expected deviation of the actual fitness from its expected value. Let us denote the expected value of the fitness of string $j \in \Omega$ representing a buyer playing against a sellers' population in state $\phi^s \in S$ by $f_j^b(\phi^s)$ and the expected fitness of j representing a seller playing against a buyers' population in state ϕ^b by $f_j^s(\phi^b)$. This notation slightly differs from the notation introduced for two-population models in section 4.7 where the fitness was written as a function of both the own and the other population state. In this framework the fitness of the string does not depend on the state of the own population but only of that of the other population. Thus, in order to keep notation as simple as possible, the own population state is neglected as an argument in the fitness function here.

Before we turn to the simulation results in this model let us examine what kind of insights can be obtained from a mathematical analysis. In order to use the results of section 4.7 we have to assume that the fitness is given by a deterministic function. So, let us assume that the actual fitness always equals the expected fitness. Thus, the fitness function in the problem reads

$$f_k^b(\phi^s) = max(0, \sum_{j \in \Omega} \phi_j^s \Pi^b(p(k), p(j))) \; k \in \Omega, \; \phi^s \in S, \tag{6.16}$$

$$f_j^s(\phi^b) = max(0, \sum_{k \in \Omega} \phi_j^b \Pi^s(p(k), p(j))) \; j \in \Omega, \; \phi^b \in S, \tag{6.17}$$

In chapter 4 we have assumed that the fitness functions are positive, continuous and continuously differentiable. Here the fitness functions are not necessarily positive but only non-negative. Hence, to be precise, an arbitrary small constant term should be added to both functions. However, this is not done here, since the influence on the actual behavior of the GA would be negligible. Note further that f is indeed continuous and continuously differentiable because for all components k with $p(k) > u$ we have $f_k^b(\phi^s) = 0 \; \forall \phi^s$ and $f_k^b(\phi^s) = \sum_{j \in \Omega} \phi_j^s \Pi^b(p(k), p(j)))$ for all k with $p(k) \leq u$. Analogous for f^s.

Application of proposition 4.7.1 in this framework indicates that in the long run the asks of the sellers and the bids of the buyers will be almost uniform. This raises the question, whether there may be a gap between these values or whether the two populations always coordinate on a one-price equilibrium as described in corollary 6.3.1. This question actually has two parts. The first one is, whether one-price equilibria are stable states of the dynamic systems and the second part is whether there are stable states of the system which are no one-price equilibrium. The first, easy part of the question is answered in proposition 6.3.2.

Proposition 6.3.2. *Every uniform profile corresponding to a Nash equilibrium (p, p), $p \in (c, u)$ of the sealed bid double auction is locally asymptotically stable with respect to the dynamics of the GA for any $\chi \in (0, 1]$ and $\mu = 0$.*

Proof. The set of uniform profiles corresponding to Nash equilibria of the auction is given by

$$N = \{(e_j, e_j) \in S \times S | j \in \Omega,\ p(j) \in [c, u]\}.$$

In order to show the local stability of (e_j, e_j), $p(j) \in (c, u)$ we just have to check that any of these states is a strict equilibrium of the economic system as defined in definition 4.7.1. First, we observe that

$$f_k^b(e_j) = \Pi^b(p(k), p(j)) = \begin{cases} u - \frac{p(k)+p(j)}{2} & p(k) \geq p(j) \\ 0 & p(k) < p(j) \end{cases}.$$

Since $p(j) < u$, we obviously have

$$f_j^b(e_j) > f_k^b(e_j) \quad \forall k \neq j \in \Omega.$$

Similar arguments show

$$f_j^s(e_j) > f_k^s(e_j) \quad \forall k \neq j \in \Omega.$$

Together this shows that (e_j, e_j) is a strict equilibrium of the system. The local asymptotic stability now follows from proposition 4.7.4.

This settles the first part of the question and shows that indeed any one-price equilibrium where the price is strictly larger than the costs but strictly smaller than the utility value is locally asymptotically stable with respect to the GA dynamics. This is hardly surprising and the more interesting part of the question is, whether there are any other uniform profiles which are also locally asymptotically stable. The answer is given in the next proposition.

Proposition 6.3.3. *Let (e_j, e_k) be a locally asymptotically stable fixed point of the GA dynamics for $\chi \in (0, 1]$ and $\mu = 0$. Then $p(j) < u, p(k) > c$ and*

$$0 \leq p(j) - p(k) \leq \left(\frac{1}{2}\right)^l. \tag{6.18}$$

Proof. Assume that (e_j, e_k) is a locally asymptotically stable fixed point of the two-population GA dynamics (4.8) where population one are the buyers and population two are the sellers. First, we show that $p(j) \geq p(k)$ must hold. Assume that $p(j) < p(k)$ or $p(j) \geq u$. This implies

$$f_j^b(e_k) = 0.$$

Let $q \in \Omega$ be a string which differs from j in only one bit position. Accordingly, we have $d(j \oplus q) = 0$ and $f_q^b(e_k) \geq f_j^b(e_k)$. If $f_q^b(e_k) > f_j^b(e_k)$, then

$$\frac{d(j \oplus q)}{l-1} = 0 < \frac{1}{\chi}\left(1 - \frac{f_j^b(e_k)}{f_q^b(e_k)}\right),$$

and proposition 4.7.3 implies that (e_j, e_k) is unstable. If $f_q^b(e_k) = f_j^b(e_k)$ then proposition 4.7.3 admits no conclusion. However, we can show that (e_j, e_k) is not locally asymptotically stable be considering the state $((1-\epsilon)e_j + \epsilon e_q, e_k)$, $\epsilon > 0$. Since q and j differ only in one bit position, any crossover between strings in the buyers' population generates a pair of offsprings that is exactly the same as the parents. Thus, $\mathcal{M}^b((1-\epsilon)e_j + \epsilon e_q) = ((1-\epsilon)e_j + \epsilon e_q)$. Further $f_q^b(e_k) = f_j^b(e_k)$ implies that $\mathcal{S}^b((1-\epsilon)e_j + \epsilon e_q, e_k) = (1-\epsilon)e_j + \epsilon e_q$. Together, this shows that $\mathcal{G}((1-\epsilon)e_j + \epsilon e_q, e_k) = ((1-\epsilon)e_j + \epsilon e_q, e_k)$ for any $\epsilon > 0$. Accordingly, there is no neighbourhood around (e_j, e_k) such that any trajectory starting in this neighbourhood converges towards (e_j, e_k), and (e_j, e_k) is not locally asymptotically stable. Hence, we have established that $p(j) \geq p(k)$ and $p(j) < u$. A similar argument establishes also $p(k) > c$.

Now, let us analyze the case where $u > p(j) \geq p(k) > c$. Let $j(m)$ denote the value of the bit at position m in string $j \in \Omega$. Define $\tilde{m} = \max(m | j(m) = 1)$ and $\tilde{j} \in \Omega$ by $\tilde{j}(m) = j(m)\ \forall m \neq \tilde{m},\ \tilde{j}(\tilde{m}) = 0$. Note that at least one bit in string j must have value '1' since $p(j) > 0$. In a similar way define $\hat{m} = \max(m | k(m) = 0)$ and $\tilde{k} \in \Omega$ by $\tilde{k}(m) = k(m)\ \forall m \neq \hat{m}\ \tilde{k}(\hat{m}) = 1$. Obviously, we have $p(\tilde{j}) < p(j), p(\tilde{k}) > p(k)$ and $d(j \oplus \tilde{j}) = 0,\ d(k \oplus \tilde{k}) = 0$. If $p(\tilde{j}) \geq p(k)$ then

$$\Pi^b(p(\tilde{j}), p(k)) = u - \frac{p(\tilde{j}) + p(k)}{2} > u - \frac{p(j) + p(k)}{2} = \Pi^b(p(j), p(k))$$

and accordingly

$$f_{\tilde{j}}^b(e_k) > f_j^b(e_k).$$

Thus, one of the stability conditions given in proposition 4.7.3 is violated, and we get a contradiction. The same argument shows also that $p(\tilde{k}) \leq p(j)$ contradicts the stability of (e_j, e_k). Therefore, the stability of (e_j, e_k) implies that either $j = k$ or $p(j) > p(k)$ with

$$p(\tilde{j}) < p(k) \quad \text{and} \quad p(\tilde{k}) > p(j). \tag{6.19}$$

In the rest of the proof, we will show that $p(j) > p(k)$ together with (6.19) can only hold if $p(j) - p(k) = \frac{1}{2^l}$. In order to do this the notation is used that $j[m] = \sum_{i=1}^{m} \frac{j(i)}{2^i}$ gives the decoded value of the first m bits of string j. Let us now consider the last bit positions in j and k. If $j(l) = 1$ then $\tilde{m} = l$ and $\tilde{j}(l) = 0,\ \tilde{j}(m) = j(m)\ m = 1 \ldots l-1$. Hence $\tilde{j}[l-1] = j[l-1]$. If $k[l-1] = j[l-1]$ either $p(j) = p(k)$ or $p(\tilde{j}) = p(k)$; $k[l-1] > j[l-1]$ implies $p(k) > p(j)$ and $k[l-1] < j[l-1]$ implies $p(k) < p(\tilde{j})$ (note that any difference in lower bit positions dominates any difference in higher bit positions). Since none of these possibilities is feasible, we conclude that $j(l) = 0$. Similar arguments establish that $k(l) = 1$. Since $p(j) > p(k)$, we must have $j[l-1] > k[l-1]$. Considering the bits positioned at $l-1$, we realize that $j(l-1) = k(l-1)$ implies $j[l-2] > k[l-2]$ which either leads to a contradiction to $p(\tilde{j}) < p(k)$ (if $j(l-1) = k(l-1) = 1$) or to $p(j) < p(\tilde{k})$ (if $j(l-1) = k(l-1) = 0$). If

$j(l-1) = 1$ and $k(l-1) = 0$, we have $j[l-2] \geq k[l-2]$ due to $p(j) > p(k)$ and $j[l-2] \leq k[l-2]$ due to $p(\tilde{j}) < p(k)$. Together this yields $j[l-2] = k[l-2]$, and we have $p(j) - p(k) = \frac{1}{2^{l-1}} - \frac{1}{2^l} = \frac{1}{2^l}$. If $j(l-1) = 0,\ k(l-1) = 1$ we consider bit position $l-2$ and use the same arguments applied to position $l-1$ to establish that either $j(l-2) = 1,\ k(l-2) = 0$ or $j(l-2) = 0,\ k(l-2) = 1$. In the former case, we have $j[l-3] = k[l-3]$ and accordingly $p(j) - p(k) = \frac{1}{2^l}$. In the later case, we continue with position $l-3$ and so on. Summarizing, we see that $p(j) > p(k)$ together with (6.19) implies that there exists some $1 \leq \tilde{m} < l$ such that $j(m) = k(m)\ \forall m < \tilde{m},\ j(\tilde{m}) = 1, k(\tilde{m}) = 0$ and $j(m) = 0, k(m) = 1\ \forall m > \tilde{m}$. Hence $p(j) - p(k) = \frac{1}{2^l}$.

Although we should be aware of the fact that these analytical results are derived under the assumption of infinite populations and zero mutation rates, they still make a very strong case that the learning process considered should converge towards an almost uniform state where the bids of buyers and sellers are very close and in the interval (c, u). These insights rely crucially on the communication effect which is incorporated by the crossover operator. The communication effect prevents any non-uniform state to be stable since there is always some probability that communication between two agents using bids in P results in one of the two changing to a bid which is no element of the set of offers respectively asks present in this equilibrium. However, communication can never yield new strategies if the population is uniform. In this sense the communication effect reinforces uniformity in the population and – as we have seen in section 6.1 – can lead to the stability of uniform states even if the strings present in the population do not have the highest possible fitness in the current environment. However, proposition 6.3.3 establishes that such a situation where communication effects lead to the stability of a non-equilibrium profile can only occur in this model if the profile is extremely close to an equilibrium profile. Thus, the probability that the system gets trapped in a profile with a substantial gap between buyers and sellers prices should be negligible and the coordination on one price equilibria should be a comparably easy task for the two populations of adaptive agents.

Let us check these findings now by looking at the actual results of numerical experiments in this framework. In all the simulation results reported here the GA parameter values $n = 100$, $\chi = 1$ and $\mu = 0.001$ were used. Changing these parameter values within a reasonable range does not make any qualitative difference in the results. Each simulation lasts for $T = 2500$ generations and the agents update their actions every $m = 20$ generations. Accordingly, the genetic operators are applied 125 times in each simulation run.

In a first step the local stability of one price equilibria with respect to the GA dynamics is examined by initializing all strings in both populations with equal values. Simulations for a large number of different initial prices show that in all cases the offers and bids stay very close to the initially given trading price throughout the run. Only minor fluctuations due to mutations can be observed in these runs. Since a positive mutation rate is used in the simulations, these results are an indicator for the stability of the one price equilibria with respect to the GA dynamics.

In experiments with random initialization the long run outcome of the process differs between single simulation runs. The observations here are based on 50 runs where the costs of the sellers are $c = 0$ and the utility of the buyers is given by $u = 1$. In all runs the process converges towards a profile where all buyers and all sellers have almost identical bids[4] and where the offers of the buyers are slightly larger than the asks of the sellers. Although the level of this pair of prices varies from simulation to simulation there is a large probability that the process ends up in a state where the good is traded at a price near 0.5. Figure 6.11 shows the frequencies with which the different prices in the interval $[0.3, 0.7]$ occur as the long run outcome of the process. There is a huge peak at $p = 0.5$ and we can see that 25 of the 50 simulation runs generate a price in the interval $[0.4875, 0.5125]$. The average price in the 50 runs is $\bar{p} = 0.4809$.

An interesting parallel can be made here to a paper by Valley et al. [168] who observed in laboratory experiments a similar distribution of trading prices in one price equilibria. They consider a setup where the values of u and c are uniformly distributed in $[0, 1]$ and buyers and sellers are allowed to communicate with their trading partner before they submit their bids. In about half of the cases where $u > c$ one price equilibria were observed and the distribution of prices looks very much like figure 6.11. In particular, in 52% of the one price equilibria the trading price was in $[0.45, 0.55]$. The corresponding percentage in our simulation runs would be 58%. Thus, it could be argued that the iterated interaction between buyers and sellers (without communication) has very similar effects than preplay communication in one shot games.

The difference between average buyers' and sellers' bids, are shown in another frequency plot in figure 6.12. We realize that the values differ significantly between the runs but that the majority of the simulations generate pairs of bids where offers and asks differ less than 0.045. The average difference between buyers and sellers bids is $\bar{p}^b - \bar{p}^s = 0.0352$. Obviously, this value is larger than the one predicted by proposition 6.3.3. This is due to the existence of mutations which may lead to a small deviation of single agents' bids from the population average. Thus, bidding exactly the opponents' average population bid might lead to the loss of some trades and is risky. In this sense the theoretical results are not supposed to give quantitative predictions about the size of deviation between buyers' and sellers' bids but rather point out the qualitative insight that no states which are qualitatively different from one price equilibria are stable with respect to the GA dynamics.

[4] I checked the variance of the prices in both populations to verify that the populations are indeed almost uniform.

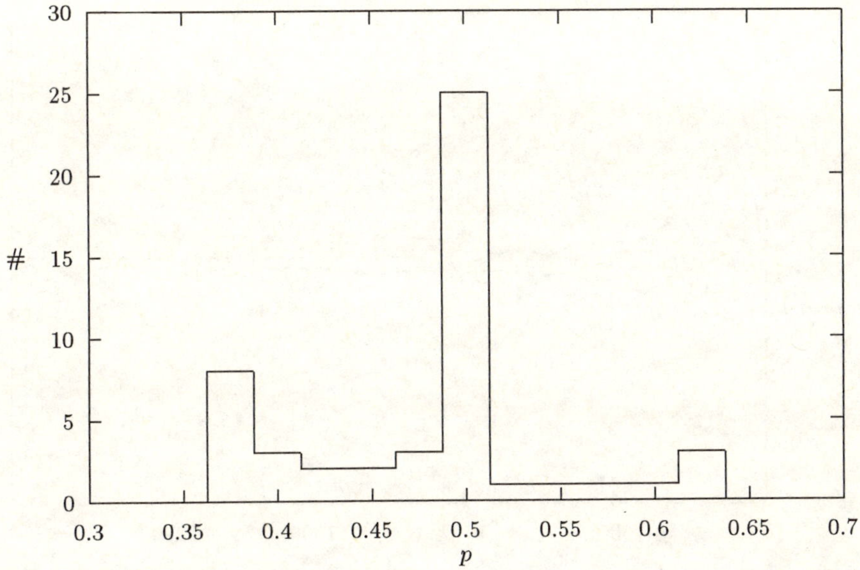

Fig. 6.11. The empirical distribution of the long run trading prices. The price range [0.3125, 0.6625] is partitioned into intervals of length 0.025 and the number of simulation runs where the price converged towards a price in the corresponding interval is given on the ordinate ($c = 0,\ u = 1,\ n = 100,\ \chi = 1,\ \mu = 0.001,\ T = 2500,\ m = 20$, 50 simulation runs).

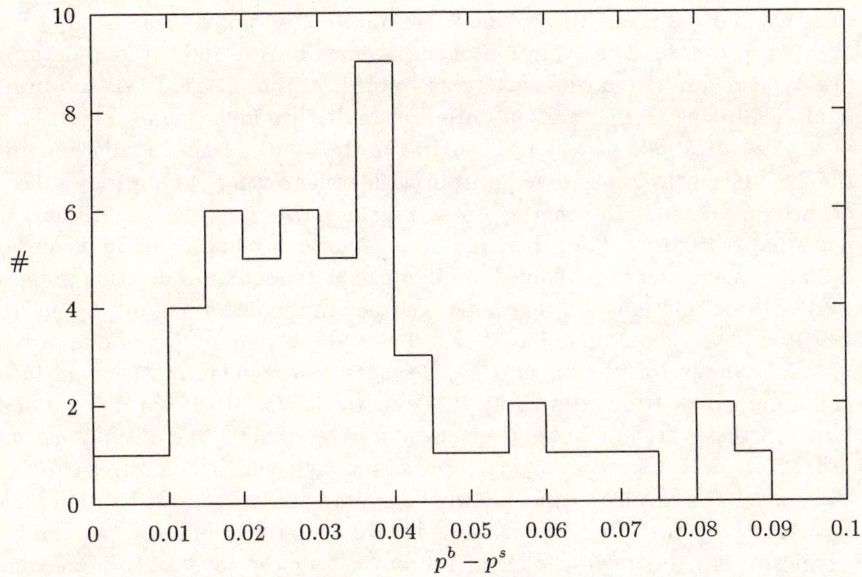

Fig. 6.12. The empirical distribution of the difference between the long run bids of buyers and sellers ($c = 0,\ u = 1,\ n = 100,\ \chi = 1,\ \mu = 0.001,\ T = 2500,\ m = 20$, 50 simulation runs).

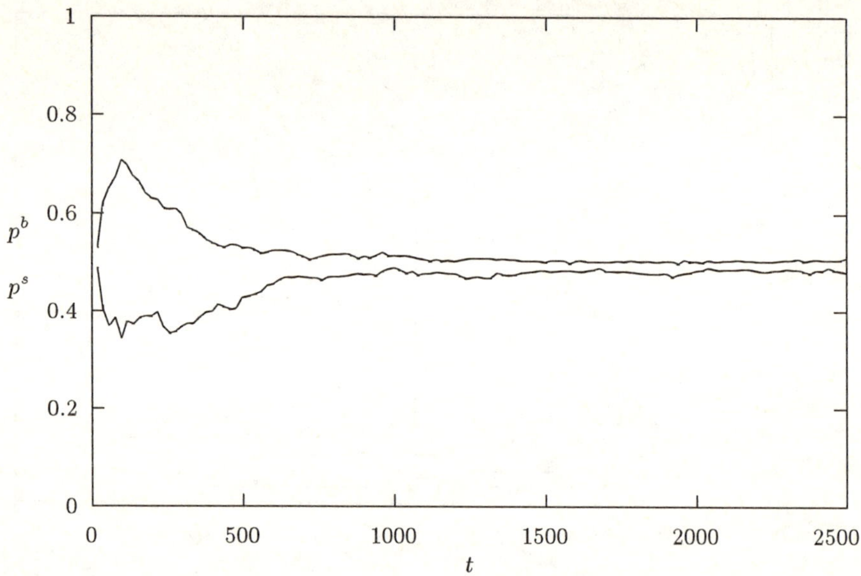

Fig. 6.13. The average population values of offers and asks in a typical simulation run ($c = 0,\ u = 1,\ n = 100,\ \chi = 1,\ \mu = 0.001,\ T = 2500,\ m = 20$).

So far we have only considered the long run outcome of the simulation runs. Of course it is also interesting to see how the two populations of adaptive agents coordinate. The evolution of the average bids of the two populations in a typical simulation run converging towards a price of $p \approx 0.5$ is shown in figure 6.13. Due to the random initialization both population averages start at a price of about $p = 0.5$. Then in the first part of the simulation run the population averages diverge from each other rather rapidly but subsequently approach each other again and settle down at values slightly larger and smaller than $p = 0.5$. This kind of behavior can be observed in all simulation runs and can be explained as follows. The random initialization implies that both populations are very heterogeneous in the beginning of the run. In fact, the payoff of a single buyer or seller in the beginning may be approximated by the payoff of a buyer or a seller against a population where the bids are uniformly distributed in $[0, 1]$. It is easy to show that the optimal bid of a buyer in these circumstances is given by $p^b = \frac{2}{3}$ whereas the optimal asking price of the seller is $p^s = \frac{1}{3}$. This implies that the agents using prices close to these bids receive a high fitness in the beginning and are imitated with a high probability by the other agents. This leads to relatively uniform population profiles where the average offer of the buyers is close to $\frac{2}{3}$ and the average ask of the sellers is close to $\frac{1}{3}$. However, in this situation it is profitable for the buyers to lower their offers and for the sellers to increase asks since the danger of loosing a trade by doing so is low. This is done in small steps due to mutation and crossover. Both populations stay almost uniform and approach each other gradually until they end up at prices close to $p = 0.5$.

In the simulations described above the learning process had a high tendency to converge towards a state where the utility gained by trading one unit of the good from a seller to a buyer is split equally between the two trading partners. One the other hand, the price $p = 0.5$ is not only the median value of $[c, u]$ but also the median value of the interval of possible bids $[0, 1]$. Thus, we should also analyze the behavior of the process in cases where these two values do not coincide. To do this, let us now deal with case where $c = 0.1$ and $u = 0.7$. The average long run trading price of 50 simulations runs with these parameter values is $\bar{p} = 0.4168$ and thus approximates the median of $[c, u]$ rather than that of $[0, 1]$. An explanation for this can be found by using arguments similar to those used above. If $c > 0$ and $u < 1$ there will be some strings in the randomly initialized population which encode bids lying below c or above u. All of these strings receive a payoff of 0 and will with probability one not be selected by any agent the first time the genetic operators are applied. Thus, a seller faces a population whose bids are uniformly distributed in $[0, u]$ rather than in $[0, 1]$ in the early stage of the simulation[5] and a buyer is exposed to bids uniformly distributed in $[c, 1]$. Calculating the best responses of both players under these assumptions we get $p^b = \frac{2}{3}u + \frac{1}{3}c$ and $p^s = \frac{1}{3}u + \frac{2}{3}c$. If we assume further that the two populations first jump to uniform states close to these values and then approach each other again we might expect that they will end up in the middle between these two values, namely near $p = \frac{u+c}{2}$. In figure 6.14 a frequency plot of the observed long run trading prices in the 50 simulation runs is depicted. The differences between offers and asks show a very similar distribution than the one shown in figure 6.12 and the corresponding distribution is not depicted here.

Figure 6.14 shows that a large number of simulations ends up with prices close to $\frac{u+c}{2} = 0.4$, however there is also a considerable peak near $p = 0.5$. There might be a rather technical explanation for this second peak. If in the first periods, where high offers yield comparably high payoffs, only offers larger than 0.5 survive in the buyers population, the buyers have a very small probability off rediscovering offers in the range $[0.25, 0.5]$ in subsequent periods when these offers are profitable (they need simultaneous mutations of the first two bids in their string). Thus, it is very likely that the sellers increase their ask towards 0.5 (they can do this with single mutation steps) before the buyers lower their offers and the resulting long run trading price is close to $p = 0.5$.

[5] Of course this does not hold in the periods preceding the first application of the genetic operators. On the other hand, the distribution of the bids is not really uniformly distributed in the periods succeeding the first application of the genetic operators, since selection has already been applied once. This shows that this reasoning is no exact description of what happens but rather an attempt to understand the basic mechanism generating the results.

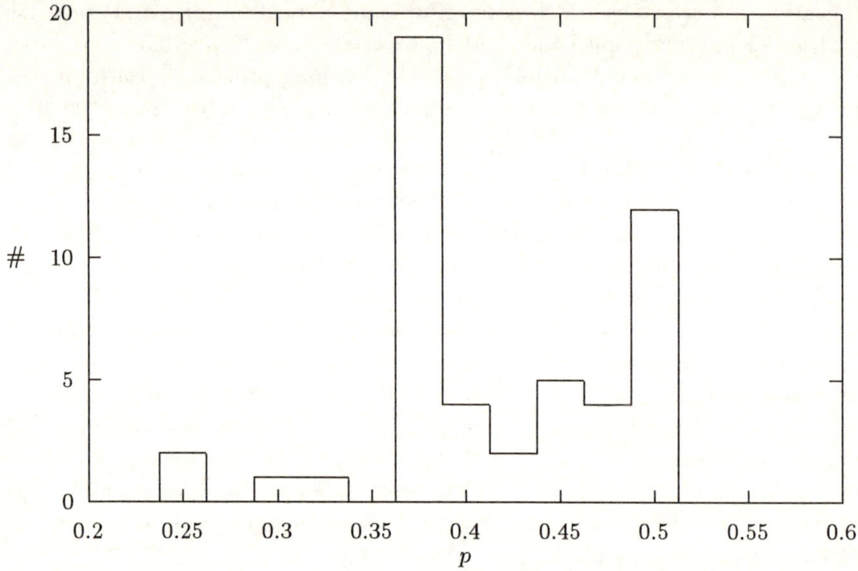

Fig. 6.14. The empirical distribution of the long run trading prices. ($c = 0.1$, $u = 0.7$, $n = 100$, $\chi = 1$, $\mu = 0.001$, $T = 2500$, $m = 20$, 50 simulation runs).

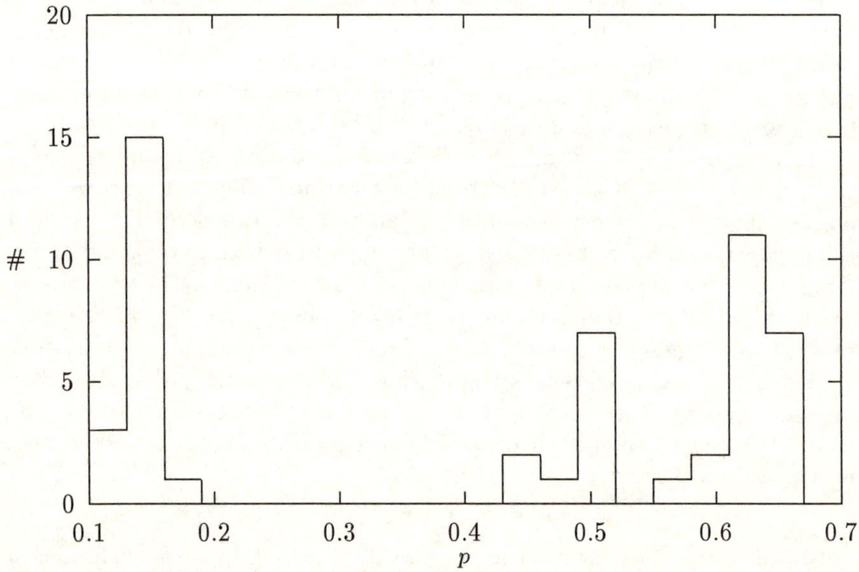

Fig. 6.15. The empirical distribution of the long run trading prices where the sellers initial asks are uniformly initialized with $p^s = 0.1$ and the buyers offers with $p^b = 0.7$ ($c = 0$, $u = 1$, $n = 100$, $\chi = 1$, $\mu = 0.001$, $T = 2500$, $m = 20$, 50 simulation runs).

The phenomenon that a lack of certain genetic material may substantially disrupt the learning of the one price equilibrium where the gain of trade is equally split, can be made much clearer if we again consider uniform deterministic initializations rather than random ones. Figure 6.15 shows the frequency distribution of long run trading prices for a set of simulations where $u = 1, c = 0$, all offers are initialized as $p^b = 0.7$ and all seller bids as $p^s = 0.1$. It can be clearly seen here that in most runs the agents' finally agree on a trading price close to one of the two initial bids and only very seldom the gain is equally split between buyers and sellers. This should hardly come as a surprise. If both populations are initialized uniformly the population where the first 'major mutation' in the right direction occurs ends up with a significantly higher profit. The new bid created by the mutation yields a much higher payoff than the rest of the population and is quickly adopted by a large part of the agents. The agents in the other population cannot change their bids significantly in subsequent periods without loosing a large number of trades and are stuck with bids close to their initial values. Only if both populations start moving their bids in the right direction approximately at the same time can the intermediate one-price equilibrium be reached. This shows that initial diversity of genetic material is essential for the selection of the 'fair' equilibrium.

Summarizing the results for deterministic and uniform cost and utility values, we might say that the simulation results back the theoretical analysis. If c and u stay constant throughout the whole period of interaction, the two population GA always converges towards a profile which closely resembles an equilibrium of the double auction market. However, due to the inherently stochastic setup, the equilibrium selected varies from run to run. Nevertheless, there is a high tendency towards the equilibria where the gain of the trade is equally divided between buyer and seller, if both populations are initialized randomly.

6.3.2 Stochastic Cost and Utility Values

Let us now turn to the case which is covered in the larger part of the literature dedicated to sealed bid double auctions, namely a situation where cost and utility values of the single agents are independent stochastic variables. These values are determined every period where their distribution is uniform in $[0, 1]$, i.e. $u_i^t \sim U[0, 1]$ *i.i.d.* $\forall i, \forall t$ and $c_i^t \sim U[0, 1]$ *i.i.d.* $\forall i, \forall t$. In this setup the strategy of an agent cannot simply state a bidding price but rather has to be a function determining the bid in dependence of u_i^t respectively c_j^t. Obviously, there is a large number of functional forms to consider and most probably there are infinitely many different equilibria for different classes of bidding functions. However, the case to which most attention has been paid is the one where the bids of the agents are linear functions of the value of u_i^t respectively c_j^t. Taking into account that a buyer never offers more than his utility value and a seller never demands less than his costs the strategies of buyers and sellers have the form

$$\pi^b(u;\alpha) = \min[u, \alpha_1 u + \alpha_2] \tag{6.20}$$
$$\pi^s(c;\beta) = \max[c, \beta_1 c + \beta_2]. \tag{6.21}$$

Chaterjee and Samuelson [38] have shown that a double auction of this type which is played between a single buyer and a single seller always has an equilibrium in linear strategies. In our framework the equilibrium values are

$$\alpha_1^* = \frac{2}{3},\ \alpha_2^* = \frac{1}{12},\ \beta_1^* = \frac{2}{3},\ \beta_2^* = \frac{1}{4}.$$

This implies of course that a population profile where all buyers use α^* and all sellers use β^* is a uniform equilibrium in the double auction market. There could be additional mixed equilibria where different agents use different linear strategies but this possibility is not further investigated here since we know that the GA in general converges towards a uniform profile. In the uniform equilibrium given above buyers and sellers make the same profit of

$$\Pi^{b*} = \Pi^{s*} = 0.0703.$$

In the GA simulations every string in the buyers population encodes two real values in $[0,1]$ interpreted as the agents choice of α_1 and α_2; analogous for the sellers population. The bids of all agents are calculated every period using formulas (6.20) and (6.21). The rest of the algorithm works exactly like in the case of deterministic values of u and c.

The function giving the expected fitness of a certain bidding strategy against a given population state is in this case considerably more complicated than in the case of deterministic, uniform cost and utility values. We have

$$f_k^b(\phi^s) = \sum_{j\in\Omega} \phi_j^s \int_0^1 \int_0^1 \Pi^b(\pi^b(u;\alpha(k)), \pi^s(c;\beta(j)))du\ ds \quad k \in \Omega, \phi^s \in S,$$

$$f_j^s(\phi^b) = \sum_{k\in\Omega} \phi_k^b \int_0^1 \int_0^1 \Pi^s(\pi^b(u;\alpha(k)), \pi^s(c;\beta(j)))du\ ds \quad j \in \Omega, \phi^b \in S,$$

where $\alpha(k)$ and $\beta(j)$ denote the corresponding parameter pairs encoded by $k \in \Omega$ respectively $j \in \Omega$. The structure of this fitness function is rather intricate if written down explicitly and so a theoretical analysis comparable to that of the last section does not seem to be feasible. A trivial insight following from proposition 4.7.4 is that the uniform profile corresponding to the equilibrium in linear strategies given above is locally asymptotically stable with respect to the GA dynamics. However, it can not be inferred whether there are other locally stable uniform profiles and also the global dynamical properties remain unknown. These questions have to be addressed with the means of numerical experiments.

Again, 50 runs were performed in this setup with parameter values

$$n = 100,\ \chi = 1,\ \mu = 0.001,\ m = 40, T = 5000.$$

The number of periods between two applications of the genetic operators has been increased since the stochastic values of u and c imply that the variance in the average payoff of an agent increases. Averaging over more periods should neutralize parts of this effect. In figure 6.16 the long run outcomes of all 50 runs are depicted. First, let me point out that also in this model the GA always converges towards an almost uniform profile, where the heterogeneity is entirely due to disruptions by mutations. The limit values of $(\alpha_1, \alpha_2, \beta_1, \beta_2)$ vary significantly between the runs. Both in the $\alpha_1 - \alpha_2$ and $\beta_1 - \beta_2$ diagram we can see a rather wide spread cloud of limit points of the different runs, where most of these points are not even close to the equilibrium values. At first sight this seems to show that the GA shows very poor learning behavior in this setup, but if we consider the form of the expected utility function of buyers and sellers given that the other part plays the equilibrium strategy (see figure 6.17) we realize that the strategies generated by the GA all lie on a very flat 'plateau' of the expected utility function and accordingly generate payoffs which are very close to that of the optimal strategies. Of course it could be argued that figure 6.16 shows that in general even in the long run the opponent does not behave according to the equilibrium strategies, but numerical calculations indicate that the shapes of these functions do not change qualitatively as long as the opponent stays on his 'plateau'. In most cases the agents expect more than 95% of the highest possible expected payoff against the other populations' strategy.

In my opinion these results can be very nicely interpreted in the spirit of a boundedly rational learning approach. The agents often do not learn the optimal strategy and may seem to be completely 'off' in a sense that their strategy parameters vary significantly form those in equilibrium. Nevertheless, due to the special structure of the model their payoff is very close to the optimal one. Thus, this looks like an example of satisficing behavior. As soon as an individual is on the utility plateau the possible gains in payoffs are so small that he stops searching and sticks to its very good but not optimal strategy. So, although these result may be disappointing if one is primarily interested in showing that GA agents are able to learn the actual equilibrium strategies, they are quite interesting and encouraging if the more general question is aksed whether boundedly rational agents interacting in such a market are able to organize in a way such that no significant losses for the traders due to coordination failures occur.

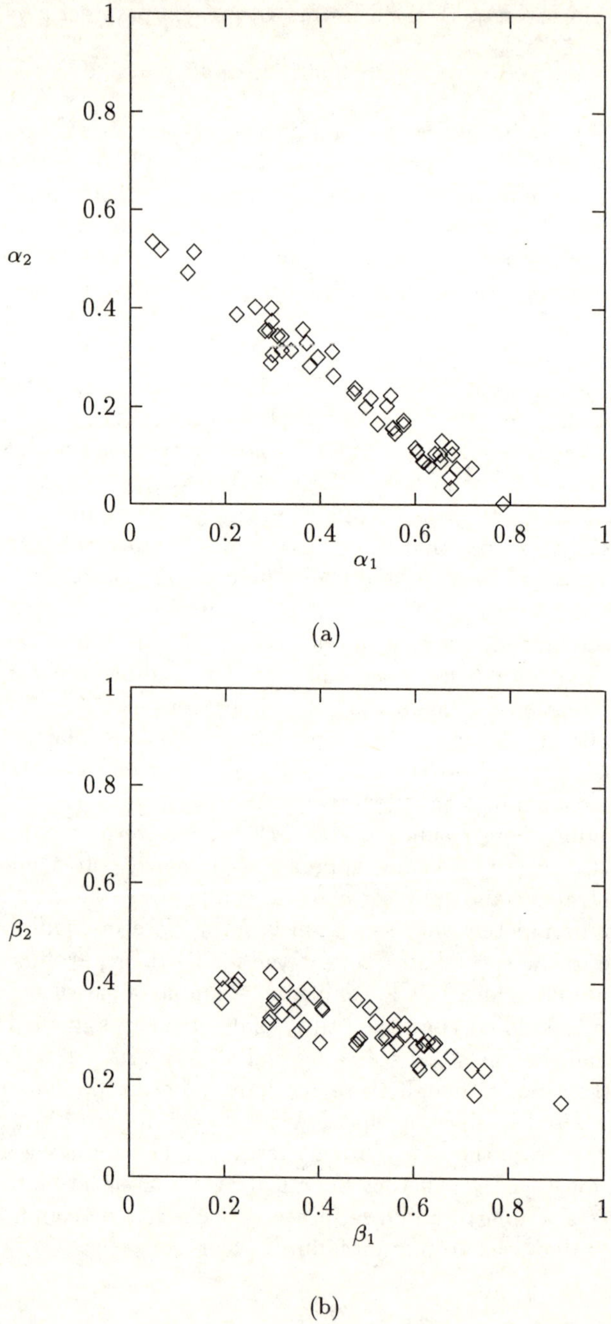

Fig. 6.16. The long run outcomes of the buyers' (a) and sellers' (b) strategy parameters in 50 different simulation runs with stochastic cost and utility values ($n = 100$, $\chi = 1$, $\mu = 0.001$, $T = 5000$, $m = 40$).

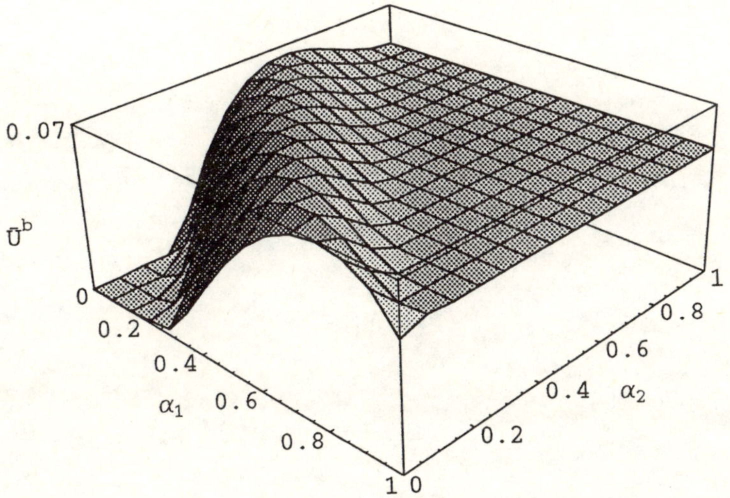

(a)

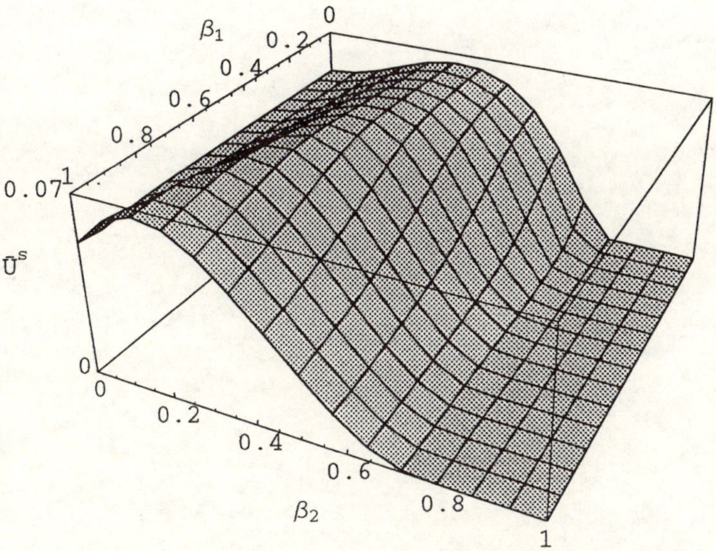

(b)

Fig. 6.17. Buyers' (a) and sellers' (b) expected payoff of an interaction with an opponent using the equilibrium strategy in dependence of the own strategy parameters.

7. Stability and Encoding

The main purpose of these notes is to demonstrate that genetic learning is a plausible model of interaction between economic agents and to study the long term properties of this learning process. The analytical results from chapter 4 have given us some insights here and keeping in mind the behavioral interpretation of the considered process the obtained insights provide information about adaptive learning per se. However, besides these economic implications the results of chapter 4 may also be very useful for the appropriate shaping of the algorithm. In this chapter I will concentrate on this aspect of genetic learning. We will see that a mathematical analysis may also be of great help for anyone who actually implements a simulation of an economic system with a GA. In particular I will propose a method which facilitates the learning of economic equilibria in a model. This may be of great importance if simulations are carried out in models where the equilibria are not a priori known. In such models a GA may be a useful tool to determine equilibria and a technique which facilitates the reaching of equilibria is of great importance.

7.1 The Cobweb Example Revisited

Consider again the cobweb model with fixed costs presented in section 6.1. Assume first that the parameters of the model are again given by (6.6) and every string in the population encodes the output value $y_{i,t} \in [0, \frac{a}{b}]$ of some firm i. We have seen in figure 6.1 that the trajectory of the average output in the population converges towards the unique homogeneous rational expectations equilibrium $y^* = \frac{5}{7}$. Thus the uniform state e_i with $y(i) \approx y^*$ is at least locally asymptotically stable. According to proposition 4.6.2, under our setup the sufficient condition for local asymptotic stability is that for all $k \neq i$

$$\begin{aligned} &\frac{d(i \oplus k)}{l-1} > \left(1 - \frac{f_i(e_i)}{f_k(e_i)}\right) \\ \Leftrightarrow\ &\frac{l-1}{l-1-d(i \oplus k)} > \frac{f_k(e_i)}{f_i(e_i)} \\ \Leftrightarrow\ &\frac{l-1}{l-1-d(i \oplus k)} > \frac{\Pi(p(e_i), y(k)) + C}{\Pi(p(e_i), y(i)) + C} \end{aligned}$$

holds. Note that the right hand side of the last expression does not depend explicitly on the string k but just on the value $y(k)$. Thus, for fixed $i \in \Omega$ we may depict the right hand side as a continuous function of $y \in \left[0, \frac{a}{b}\right]$. In order to analyze the stability properties of e_i we just have to plot all the values of $\frac{l-1}{l-1-d(i \oplus k)}$ into the same figure as this continuous function and check whether all points are above the graph of the function.

The corresponding plot for the homogeneous rational expectations equilibrium of the cobweb model is shown in figure 7.1. The continuous line is the graph of the function $\frac{1}{\Pi(p(e_i),e_i)+C}(\Pi(p(e_i), y) + C)$, where $i = 1011011011$ encodes the equilibrium value y^*. Each of the diamonds in the picture corresponds to one string $k \in \Omega$. On the abscissa the output value encoded by the string is given and on the ordinate the inverse of the probability that the string is not disrupted by crossover with string i. All points are above the graph of the function which implies that the uniform state e_i is locally asymptotically stable. Of course we knew this already, but this graphical method allows us a rather simple stability check.

However, it is more convenient to use a slightly different plot for the graphical stability check. To plot $y(k)$ on the ordinate is no natural way to present this plot because $y(k)$ is a variable in the model but has no intrinsic relevance to the GA. Relevant for the stability of the state e_i is how large $d(i \oplus k)$ is compared to the fitness ratio. Thus it seems to be natural to put these two values on ordinate and abscissa. In figure 7.2 we see the corresponding plot. Again, the string i is the one encoding the equilibrium output y^*. The continuous line is the graph of the function $s(d) = \frac{l-1}{l-1-d}$. Again each diamond represents one string $k \in \Omega$. The distance $d(i \oplus k)$ of this string from the string i is the abscissa value of the diamond, the fitness ratio $\frac{f_k(e_i)}{f_i(e_i)}$ is the value on the ordinate. Whenever all points are below the function $s(d)$ the uniform state e_i is locally asymptotically stable. Obviously e_i is an economic equilibrium if and only if all abscissa values are equal to or smaller than one and a strict economic equilibrium if they are all strictly smaller than one. Note that the function $s(d)$ does not depend on any details of the model like the fitness function or the encoding mechanism but only on the parameter l of the GA. Thus $s(d)$ is some kind of universal stability barrier for all models where strings with the same length are used. Of course the cloud of points in the plot depends on the particular model used, but we always have to compare this cloud with the same function $s(d)$.

However the equilibrium state is not the only locally stable state in this system. If we plot the graphical stability check for the uniform state consisting only of strings $i = 1100000000$ corresponding to $y = 0.75$ we get figure 7.3. It can be seen quite clearly in this figure that also this state is locally asymptotically stable. However, the basin of attraction of this stable point seems to be much smaller than the one of the equilibrium state as the trajectory converged towards the equilibrium in all our simulations. This shows that local asymptotic stability does of course not imply that this state will actually be reached by the algorithm. These two states are the only stable states in the system. To present also a stability diagram for an unstable

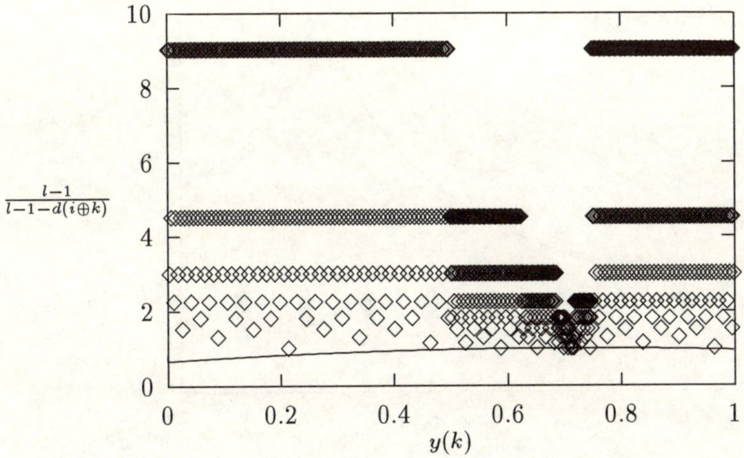

Fig. 7.1. A graphic illustration of the stability criterion for the rational expectations equilibrium in the cobweb model

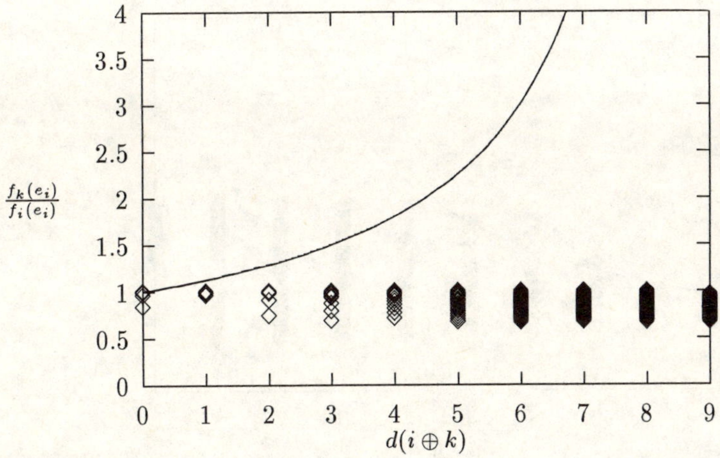

Fig. 7.2. Graphic stability criterion for the rational expectations equilibrium in the cobweb model

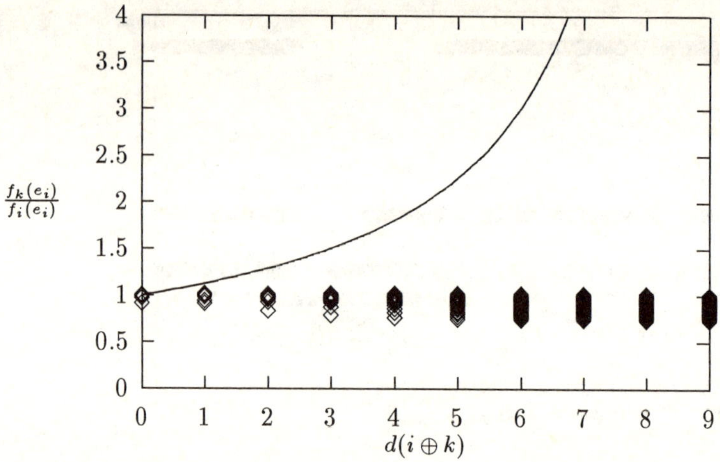

Fig. 7.3. Graphical stability check for the uniform population with average production of $\bar{y} = 0.75$

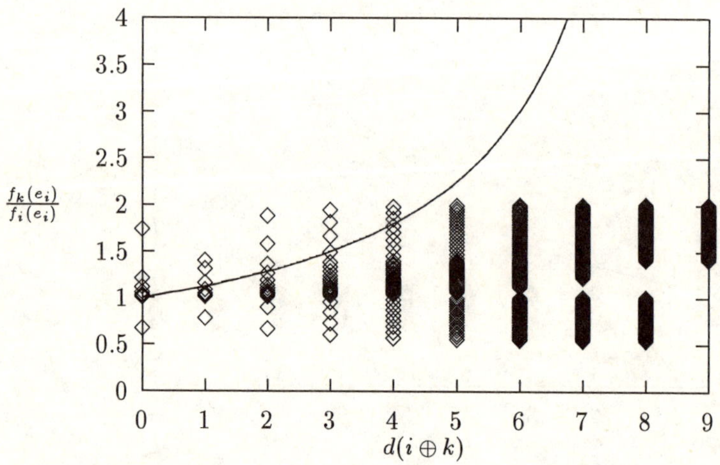

Fig. 7.4. Graphical stability check for the uniform population with average production of $\bar{y} = 0.25$

state, the corresponding plot for the uniform state e_i with $i = 0100000000$ is depicted in figure 7.4. The production decision encoded by this string is to produce an amount of $y = 0.25$. This state is of course no equilibrium and the figure shows that it is also unstable. A large number of points lie above the stability barrier, each one indicating an eigenvalue larger than one.

7.2 Impact of a Change in Encoding and Scaling

We have seen in the last section that the graphical stability check may be a very useful tool to determine the stability properties of uniform states. In order to show the impacts of a change in the encoding mechanism I again use the example of the cobweb model. However, this time I consider the case of higher fixed costs given by (6.7). Recall that the old homogeneous rational expectations equilibrium is no economic equilibrium any more but that it is locally stable if pure quantity decisions are considered. Considering again the stability criterion for the uniform state e_i with $i = 1011011011$ (figure 7.5) we realize that there is only a small difference to the case with the lower fixed costs. Only one point with distance 9 from string i has moved from a fitness ratio smaller than one to a fitness ratio larger than one. Nevertheless, all points are below the stability barrier $s(d)$. This shows again that e_i is no economic equilibrium but locally asymptotically stable for the GA. However, we know from our simulations that this state becomes unstable if we change the encoding mechanism by adding one special bit determining the market exit or entry decision of the firm. The changes in our graphical stability check can be seen in figure 7.6. The single difference to figure 7.5 is that the point with ordinate value larger than one has multiplied and appears now for any value of $d(i \oplus k)$. Thus, three points are above the stability barrier and the state is unstable now.

In general a change of the encoding mechanism may change the distance between two strings. Thus a change of coding may shift any point horizontal in our graph. Any stable state which is no economic equilibrium may be made unstable if the coding is altered in a way that the points with fitness ratio larger than one shift sufficiently to the left. Due to proposition 4.6.5 we know that this is always possible. Note that also the inverse is possible. By shifting all points lying above the stability frontier to the right an unstable fixed point may become stable for a GA with changed coding.

Another possibility to alter the properties of the graphical stability check is to change the length of the strings. A way to do so would be the adding of genetic waste bits (see section 4.6). This increases l and rotates the function determining the stability barrier downwards towards the horizontal line with ordinate value 1. If the addition of the genetic waste bits does not change the distance values $d(i \oplus k)$ (in particular if all genetic waste bits are added at the beginning or the end of the string) this will eventually lead to the instability of any non-equilibrium state. If the distance values change too the effect is

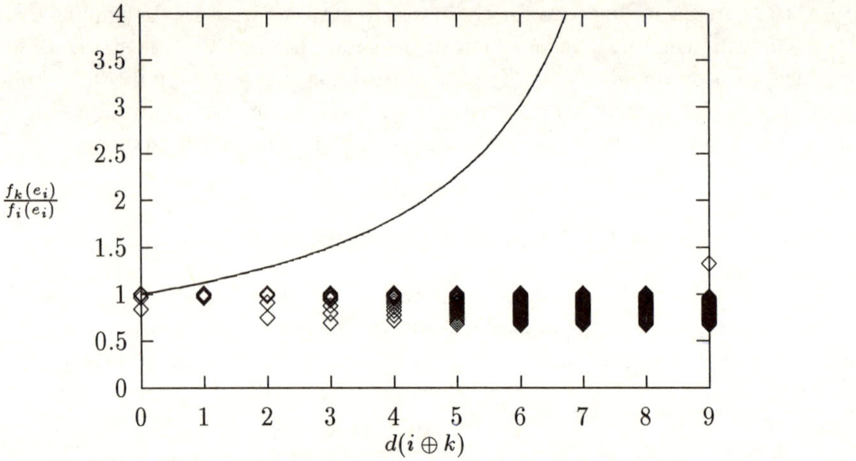

Fig. 7.5. Graphical stability check for the "false" homogeneous equilibrium state in the case of pure quantity decisions

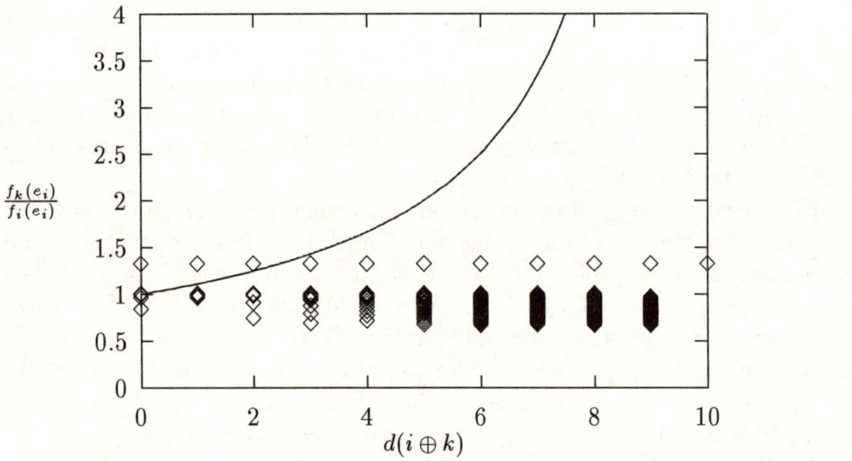

Fig. 7.6. Graphical stability check for the "false" homogeneous equilibrium state in the case of separated exit and entry decisions

not so clear. In such a case the addition of genetic waste may even have a stabilizing effect.

Note also that the stability diagram shows whether a rescaling of the fitness function may lead to a change of the stability properties. As any scaling has to be a monotone increasing transformation, any fitness ratio smaller than one will still be smaller than one after the rescaling. Thus we get as the first trivial insight that an economic equilibrium is always stable regardless of the fitness scaling. However it is also quite clear that any fitness ratio larger than one may be made arbitrarily large by a monotone increasing transformation. This can be done for example by taking the old fitness to some high power. This implies that almost any non-equilibrium state may be made unstable by rescaling the fitness. To see that this does not work for any such state consider figure 7.5. The only string k where $\frac{f_k(e_i)}{f_i(e_i)} > 1$ holds has distance $l-1$ from string i. However the value of $s(d)$ converges towards infinity for $d \to l-1$. Thus, no rescaling can cause the fitness ratio to reach the stability barrier. This implies that e_i can not be made unstable by rescaling. Similar arguments apply also for the variation of the crossover probability χ. Throughout this chapter it was always assumed that $\chi = 1$ but it gets quite obvious from proposition 4.6.2 that also a variation of χ may change the stability properties of a uniform non-equilibrium state. Summarizing we realize that an asymptotically stable uniform state e_i which is no economic equilibrium may be made unstable by changing the encoding scheme, changing certain parameter values or rescaling the fitness (if there is a string with a higher fitness than i at e_i which does not differ from i in the first and the last bit). Remember, however, that our theoretical results were derived for the stylized case of an infinite population. If we consider a population of given finite size any non-equilibrium state will in practice become unstable if the corresponding fitness ratio is large enough.

7.3 A Method for Finding Economic Equilibria

The arguments given above and the graphical stability analysis may be used to give some recommendations as how to adapt the GA to find a uniform economic equilibrium in a system. Assume that we have carried out a simulation with a GA in an economic system where we do not know the equilibria. Assume further that we are not mainly interested in the study of the behavior of the GA per se but rather in finding an economic equilibrium of the model. Our process has converged to some uniform state and, of course, we are interested in determining whether or not this state is an equilibrium.

We know from proposition 4.6.5 that this state will be stable for any coding mechanism if it is an economic equilibrium but that it becomes unstable for some coding if it is no economic equilibrium. Thus, we could just change the coding in an arbitrary fashion for several times and see whether the process stays near the uniform state or converges to some other uniform state. In the first case we may conclude that with some probability depending on the number of trials our state is indeed an economic equilibrium. In the second case we know that this state is no equilibrium and have to apply the same procedure for the state the GA has now converged to. Although cycles may appear we can hope to arrive at an economic equilibrium sooner or later.

All the same, randomly generated changes in the encoding mechanism are not very efficient. Using our stability diagram we may determine exactly which changes in the coding have to be made in order to make a stable non-equilibrium state unstable. Thus, when a GA has converged to some uniform state e_i we should start to calculate the fitness ratio and the distance from i for strings $k \neq i$ and plot them into the stability diagram. If all fitness ratios are smaller than one e_i is an economic equilibrium. If e_i is no equilibrium we might stop as soon as we have found the first point with fitness ratio larger than one. We change the coding in such a way that this point shifts to the left above the stability frontier $s(d)$. Then we carry out a new simulation with the GA with this new coding. As e_i is unstable now the process will settle down at some different uniform state e_j. Calculating again the points in the graphical stability check we either see that this point is an equilibrium or realize how to change the encoding mechanism to make also this uniform state unstable. Keeping track of all the uniform states the algorithm has converged to before, we may always try to change the coding in such a way that the uniform states which were made unstable before do not become stable again due to subsequent changes. If this is possible no cycles may appear and we have to arrive at the equilibrium after some steps if a uniform equilibrium exists in the system. However, it is not clear to us whether it is always possible to find the appropriate coding mechanism which avoids convergence to all non-equilibrium states.

The method described above gives a simple recommendation as how to adapt the GA to find uniform economic equilibria in complex economic systems. The computational effort to carry out this method may be rather large if Ω is large, but it should be pointed out that in general the fitness values $f_k(e_i)$ have to be calculated only for a subset of Ω. Of course, a more rigid mathematical analysis of this algorithm could be carried out but I do not intend to do this here. I just wanted to show in this section how analytical results may be used to facilitate the finding of economic equilibria. If the payoff function of the system is a black box the effort of the explicit calculation of an economic equilibrium is immense as the fitness values $f_k(\phi)$ for all states $\phi \in S$ and all strings $k \in \Omega$ have to be calculated. In such a case the approach suggested here may be an attractive alternative.

8. Conclusions

These notes are intended to contribute to the growing field of research dealing with the use of AI and CI techniques in economics. I have shortly summarized some of the most important contributions of the recent years and have afterwards concentrated on a thorough analysis of a special CI technique, namely genetic algorithms. It was argued that GAs, due to their decentralized structure which very naturally resembles to a group of economic agents and their interactions, are especially well-suited to simulate the behavior of an economic system. Further an interpretation in a behavioral sense of the single operators contained in a genetic algorithm was suggested, but also the problems with the economic interpretation of certain aspects of the algorithm were pointed out. The fact that in economic setups the fitness of a single string depends on the current state of the whole population has led to the conclusion that the analytical models which describe the behavior of a genetic algorithm used to solve an optimization problem can not be applied in most economic systems. Perhaps the main results are several propositions describing the limit behavior of a genetic algorithm in systems where the fitness function is state dependent. These theoretical results, but of course also the learning ability of a GA as such, have been illustrated afterwards with several examples from the field of game theory and economics. The simulations showed that mathematical reasoning enables us not only to understand but also to predict the behavior of GAs in economic systems. However these examples made also clear that a local analysis like the one done in chapter 4 is in many cases not able to explain the complete behavior of the system. In several cases heuristic explanations for a certain kind of behavior had to be given instead of a rigorous mathematical analysis. We have to face the fact that the current knowledge about high-dimensional non-linear difference equations does not enable us to describe the global behavior of such a system. Nevertheless, I am confident that the application of special theories like the graph theoretical approach taken here in proposition 4.2.1 or the theory of nonlinear dynamical systems will permit further mathematical insights into the behavior of GAs in SDF systems.

Recalling the arguments in section 2.5 in favor of the use of genetic algorithms instead of analytic learning models, it seems perhaps a little odd to derive an analytical representation of the GAs at all. One of the main advantages of simulation techniques like genetic algorithms is that they may be applied in systems which are analytically not tractable. However in analyzing the behavior of the GA the main goal is not to completely describe the exact behavior of the GA, but rather to gain some insight into the effects of certain parameter constellations, coding mechanisms or applications of different genetic operators. The knowledge of which behavior is basically possible and which is not is important for two reasons. First, we get information about the connection between behavioral assumptions about single individuals and the learning ability of the whole population, and second, it is of great help to choose the best suited kind of GA for a certain simulation. This was shown quite clearly in the last chapter. The analytical considerations can never replace the simulations, but may be a useful tool to facilitate the building of a well-working GA and the understanding of the obtained results.

Considering our results from an economists' point of view a main conclusion may be that we have no guarantee that an economic system where the individuals adapt their strategies by imitation, exchange of information and innovation will in the long run arrive at an equilibrium state. Both the analytical, but also the numerical evidence says that there are situations where a genetic algorithm converges towards a non-equilibrium state. A negative answer to the basic question whether adaptation necessarily leads to equilibrium behavior implies however that it is even more important to understand the behavior of the system out of equilibrium. Probably the most convenient and also most accessible way to analyze this behavior is to use a model which may be implemented on a computer and simulate the behavior. The fast growing agent-based computational economics community is going in this direction. It is hard to estimate the impact this kind of research will have on economic theory, because compared to other fields of economic research the work in this direction has just begun. I believe that this approach may indeed become a major part of economic research and may offer a new understanding of economic behavior. However, for this project to be successful it will be important to have a sound theoretical foundation and a good linkage to empirical evidence. The intention of this monograph was to provide parts of the theoretical foundation.

A. Basic Definitions and Results Used

A.1 Time Homogeneous Markov Chains

Definition A.1.1. *Let S be the state space of the stochastic process in discrete time $\xi = \{\xi_t,\ t = 0, 1, \ldots\}$. The process ξ is called a Markov chain on S iff the following condition is satisfied:*

$$\mathbb{P}(\xi_{t+1} = j | \xi_0 = i_0, \xi_1 = i_1, \ldots, \xi_t = i_t) = \mathbb{P}(\xi_{t+1} = j | \xi_t = i_t)\ \ \forall i_k, j \in S.$$

The probability

$$q_{ij}(t) = \mathbb{P}(\xi_{t+1} = j | \xi_t = i)$$

is called the transition probability from i to j at time t. The matrix $Q(t) = [q_{ij}(t)]$ is called the transition matrix of the process ξ at time t.

Definition A.1.2. *A Markov chain ξ is called time homogeneous or a chain with stationary transition probabilities if the transition matrix is independent of t.*

Definition A.1.3. *Let ξ be a time homogeneous Markov chain with transition matrix Q. The state $i \in S$ is called an absorbing state, iff $q_{ii} = 1$.*

Definition A.1.4. *Let ξ be a time homogeneous Markov chain with transition matrix Q. The set $C \subseteq S$ is called a closed set, iff $q_{ij}^{(k)} = 0$ for all $i \in C$, all $j \notin C$ and all $k \geq 1$, where $q_{ij}^{(k)}$ is the probability of transition form i to j in k steps.*

Definition A.1.5. *Let ξ be a time homogeneous Markov chain with transition matrix Q. A probability distribution π over S is called a stationary distribution of ξ iff $\pi^T Q = \pi^T$.*

Definition A.1.6. *A Markov chain ξ is called irreducible if for every two states i and j there exists a positive integer k such that $q_{ij}^{(k)}$ is positive.*

Definition A.1.7. *A time homogeneous Markov chain ξ has period d, iff the following conditions hold for all $i, j \in S$:*

1. $q_{ij}^{(n)} = 0\ \forall n \neq md,\ \ m \in \mathbb{N}$

2. d is the largest integer with property 1.

If $d = 1$ the chain is called aperiodic.

Definition A.1.8. *A square matrix Q is called primitive iff there exists a positive integer k such that $Q^k > 0$.*

With these definitions we get the following trivial lemma:

Lemma A.1.1. *Every time homogeneous Markov chain with a primitive transition matrix is irreducible. In particular, every time homogeneous Markov chain with a positive transition matrix is irreducible.*

The following theorem is a standard result from the theory of time homogeneous Markov chains. A proof can be found for example in Adke and Manjunath [1], p.138:

Theorem A.1.1. *Let ξ be an irreducible aperiodic time homogeneous Markov chain on the finite state space S. Then ξ has a unique stationary distribution π. For every initial distribution π^0 holds $\lim_{t\to\infty} \left(\pi^0\right)^T Q^t = \pi^T$.*

The next theorem, which is used in the proof of proposition 4.2.1 is a non standard result, and therefore a proof is given here. The theorem was first discovered by Hill [82] in the context of certain biological models. An extension of this technique to non irreducible chains was done by Leighton and Rivest [106], [107] or Anantharam and Tsoucas [5]. The proof given here is similar to the one given by Freidlin and Wentzell [66], p.177.

Theorem 4.2.1. *Consider an irreducible Markov chain on the finite state space S with transition matrix Q, then the unique stationary distribution π is given by $\pi = (\sum_{\phi\in S} \nu_\phi)^{-1}\nu$ with*

$$\nu_\phi = \sum_{g\in G(\phi)} \prod_{(x,y)\in g} q_{xy} \quad \forall \phi \in S.$$

Proof. Due to theorem A.1.1 we know that the solution of the equation $x^T Q = x^T$ is unique up to a multiplicative constant. We have to show therefore only that $\nu^T Q = \nu^T$ or written explicitly that

$$\sum_{z\in S} \nu_z q_{z\phi} = \nu_\phi \quad \forall \phi \in S$$

Subtracting $\nu_\phi q_{\phi\phi}$ on both sides we get

$$\sum_{z\neq\phi} \nu_z q_{z\phi} = \nu_\phi(1 - q_{\phi\phi})$$

Inserting the full expressions for the ν_z and using that $\sum_{z\in S} q_{\phi z} = 1$ we get as an equivalent equation to the above:

$$\sum_{z\neq\phi}\sum_{g\in G(z)}\prod_{(x,y)\in g} q_{xy}q_{z\phi} = \sum_{u\neq\phi}\sum_{h\in G(\phi)}\prod_{(v,w)\in h} q_{vw}q_{\phi u}. \tag{A.1}$$

To complete the proof it only has to be shown that the above equality holds. Take a summand from the sum on the left hand side for an arbitrary value of $z \neq \phi$ and an arbitrary z-graph g. As $z \neq \phi$ there must be a sequence of arrows in g leading from ϕ to z. Let us call the end point of the first arrow in this sequence u. Of course u could be the state z itself, but we always have $u \neq \phi$. We construct now a ϕ-graph h by removing the arrow (ϕ, u) from g and adding the arrow (z, ϕ). It is easy to see that for these two graphs g and h

$$\prod_{(x,y)\in g} q_{xy}q_{z\phi} = \prod_{(v,w)\in h} q_{vw}q_{\phi u},$$

which means that every summand in the sum on the left hand side appears also in the sum on the right hand side. Conversely, take the summand on the right side of (A.1) for some ϕ-graph h and some state $u \neq \phi$. As $u \neq \phi$ there has to be sequence of arrows leading from u to ϕ. Let us call the initial point of the last arrow in this sequence z. We have therefore $z \neq \phi$. By removing the arrow (z, ϕ) and adding (ϕ, u) we construct a z-graph g, and again we have

$$\prod_{(x,y)\in g} q_{xy}q_{z\phi} = \prod_{(v,w)\in h} q_{vw}q_{\phi u}.$$

This implies that every summand from the sum on the right hand side of (A.1) appears also on the left hand side. Therefore (A.1) must be satisfied, and we have shown the theorem.

A.2 Nonlinear Difference Equations and Stability

Definition A.2.1. *Let $F : S \to S$ be an operator on the state space $S \subseteq \mathbb{R}^n$. A sequence $\{x_t\}_0^\infty$ with $x_t \in S$ $\forall t$ satisfying*

$$x_{t+1} = F(x_t) \;\; \forall t = 0, \ldots \tag{A.2}$$

is called a solution or a trajectory of the difference equation (A.2).

Definition A.2.2. *An element $x \in S$ is called a fixed point of the difference equation (A.2) iff*

$$x = F(x).$$

A point $x \in S$ with $x = F^k(x)$, $x \neq F^m$ $\forall m < k$ for some integer k is called a periodic point with period k of (A.2).

Definition A.2.3. *Let $x \in S$ be a fixed point of the difference equation (A.2). Then*

(i) x *is (Liapunov) stable with respect to (A.2), if for any* $\epsilon > 0$ *there is a* $\delta(\epsilon) > 0$ *with* $F^t(y) \in B(x, \delta)\ \forall y \in B(x, \epsilon(\delta)),\ \forall t,$

(ii) x *is (locally) asymptotically stable, if* x *is stable and there exists an* $\epsilon > 0$ *such that* $\lim_{t\to\infty} F^t(y) = x\ \forall y \in B(x, \epsilon),$

(iii) x *is globally asymptotically stable, if* x *is stable and* $\lim_{t\to\infty} F^t(y) = x$ *holds for all* $y \in S,$

where $B(x, \epsilon)$ *denotes the open ball in* $\mathbb{R}^n$ *with radius* ϵ.

Definition A.2.4. *Let $x \in S$ be a fixed point of (A.2). A continuous function $V : S \to \mathbb{R}_+$ is called a Liapunov function for F at x, iff for some $\epsilon > 0$ the following three conditions are satisfied:*

(i) $V(x) = 0$

(ii) $V(y) > 0\ \forall y \in B(x, \epsilon),\ y \neq x$

(iii) $V(F(y)) \leq V(y)\ \forall y \in B(x, \epsilon).$

If the inequality in (iii) is strict for all $y \neq x$, V is called a strict Liapunov function. V is a global (strict) Liapunov function, if the conditions (ii) and (iii) are satisfied on the entire set S.

A proof of the following theorems concerning the stability of fixed points can for example be found in Kelley and Peterson [94].

Theorem A.2.1. *Let x be a fixed point of (A.2), and assume that F is continuous. If there is a Liapunov function V for F at x, then x is stable with respect to (A.2). If there is a (global) strict Liapunov function, then x is (globally) asymptotically stable.*

Theorem A.2.2. *Let x be a fixed point of (A.2), and let the square matrix A be the linearization of F at x, i.e. $F(y) = x + A(y - x) + G(y)$ with $\lim_{y\to x} \frac{\|G(y)\|}{\|y-x\|} = 0$. If all eigenvalues of A lie within the interior of the unit ball in the complex plane, the fixed point x is locally asymptotically stable. If there exists an eigenvalue with absolute value larger than one the fixed point x is unstable.*

B. Calculation of the Equilibria of the Evolutionary Games in Chapter 5

In this chapter the exact calculations of the equilibria of all games considered in Chapter 5 are provided.

B.1 Rock-Scissor-Paper Games

We consider a circulant Rock-Scissors-Paper game, whose payoff matrix is given by (5.3). As we are analyzing this class of games in an evolutionary context, we are only interested in the symmetric equilibria of the game. We proceed in three steps. First, we have to check whether equilibria in pure strategies exist. This would mean, that there exists a pure strategy i such that

$$e_i^T A e_i \geq e_k^T A e_i \ \forall k \in I.$$

To keep the notation as simple as possible let us denote by $B(m)$ the set of all mixed strategies which are best replies to the mixed strategy m:

$$B(m) = \{s \in \Delta^3 | s^T Am \geq \tilde{s}^T Am \ \forall \tilde{s} \in \Delta^3\}.$$

It is easy to see that the best reply structure in RSP games is given by

$$\begin{aligned} B(e_1) &= \{e_3\} \\ B(e_2) &= \{e_1\} \\ B(e_3) &= \{e_2\}. \end{aligned}$$

Thus, there is no strategy $i \in I$ with $e_i \in B(e_i)$. Acoordingly, no symmetric equilibria in pure strategies exist.

As the second step, we have to show that there are no symmetric equilibria with a support of two pure strategies. Let us assume that there is an equilibrium strategy $m = (m_1, m_2, 0)$, $m_1, m_2 > 0$, $m_1 + m_2 = 1$. Since (m, m) is a symmetric equilibrium of the game, $e_2 \in B(m)$ has to hold. However, we have

$$e_2^T Ax = m_1 z + m_2 x < m_1 x + m_2 y = e_1^T Ax.$$

This shows that e_2 is no best reply to m and accordingly m is no equilibrium of the game. Since the payoff matrix is completely symmetric, the same argument can be applied to any strategy with a support of two pure strategies.

Thus, there is no symmetric equilibrium where only two pure strategies are used.

We know that any symmetric normal form game has at least one symmetric Nash equilibrium, which implies that there has to be a completely mixed equilibrium. Such an equilibrium is given by a vector $m = (m_1, m_2, m_3)$ satisfying

$$Am = \lambda \mathbb{1}$$

for same scalar $\lambda \in \mathbb{R}$ ($\mathbb{1} = (1,1,1)^T$). Considering $m^* = \frac{1}{3}(1,1,1)^T$ we get

$$Am^* = \frac{x+y+z}{3}\mathbb{1}$$

which shows that (m^*, m^*) is indeed a symmetric Nash equilibrium of the game. Let us assume that there is another strategy $\tilde{m} \neq m^*$ with

$$A\tilde{m} = \lambda \mathbb{1}$$

for some $\lambda \in \mathbb{R}$. In this case we have

$$A(u\tilde{m} + vm^*) = \left(u\lambda + v\frac{x+y+z}{3}\right)\mathbb{1}$$

for any $u, v \in \mathbb{R}$, which implies that any linear combination of $\tilde{m}$ and m^* which is inside the simplex Δ^3 defines a symmetric Nash equilibrium. However, two of these linear combinations must lie on the boundary of Δ^3. Since we have shown above that there are no equilibria on the boundary, the assumption that there is a second symmetric equilibrium $\tilde{m}$ leads to a contradiction. Accordingly, m^* in the unique symmetric Nash equilibrium of the game.

Finally, we show that m^* is an ESS if and only if

$$x < \frac{y+z}{2}. \tag{B.1}$$

Note that

$$m^T Am^* = {m^*}^T Am^*$$

holds for all $m \in \Delta^3$. Accordingly, m^* is an ESS if and only if

$${m^*}^T Am > m^T Am$$

is satisfied for all $m \neq m^*$. If $x \geq \frac{y+z}{2}$ this inequality is violated for $m = e_i$, $i = 1, 2, 3$. Thus, (B.1) is a necessary condition for m^* to be an ESS. In order to show that it is also sufficient we have to establish that

$$G(m) := {m^*}^T Am - m^T Am$$

is positive for all $m \in \Delta^3$ with $m \neq m^*$. Writing the payoff matrix of the game as

$$A = \tilde{A} + \hat{A},$$

with

$$\tilde{A} = \begin{pmatrix} 0 & \alpha & -\beta \\ -\beta & 0 & \alpha \\ \alpha & -\beta & 0 \end{pmatrix}, \qquad \alpha = y - x > 0,\ \beta = x - z \geq 0$$

and

$$\hat{A} = x \begin{pmatrix} 1\,1\,1 \\ 1\,1\,1 \\ 1\,1\,1 \end{pmatrix}$$

we get

$$u^T \hat{A} v = x \ \forall u, v \in \Delta^3, \qquad m^{*T} \tilde{A} m = \frac{1}{3}(\alpha - \beta) \ \forall m \in \Delta^3$$

and, accordingly,

$$G(m) = \frac{1}{3}(\alpha - \beta) - m^T \tilde{A} m = \frac{1}{3}(\alpha - \beta) - (m_1 m_2 + m_1 m_3 + m_2 m_3)(\alpha - \beta).$$

Under condition (B.1) $\alpha - \beta$ is positive and we are left with showing that $m_1 m_2 + m_1 m_3 + m_2 m_3 < \frac{1}{3} \ \forall m \in \Delta^3, m \neq m^*$. However, it is easy to realize that the left hand side of the inequality is maximal for $m_1 = m_2 = m_3 = \frac{1}{3}$. This establishes that (B.1) is also sufficient for m^* to be an ESS.

B.2 The GA Deceptive Game GAD

Now, consider the symmetric game with payoffs given by (5.4). First,we note that $e_1 \in B(e_1)$, which implies that (e_1, e_1) is a symmetric equilibrium of the game. Since all other strategies have a strictly smaller payoff against e_1 than e_1 itself the conditions of an ESS are met too. Straightforward calculations show that there is no other symmetric equilibrium in pure strategies. Next let us check, whether there are symmetric equilibria (m, m) with a support of two pure strategies. We distinguish three cases:

1.) $m = (m_1, m_2, 0),\ m_1, m_2 > 0, m_1 + m_2 = 1$
If (m, m) is a symmetric equilibrium, $e_2 \in B(m)$ has to hold. However

$$e_3^T A m > e_2^T A m$$

holds for any m with $m_3 = 0$, which shows that $e_2 \notin B(m)$.

2.) $m = (m_1, 0, m_3),\ m_1, m_3 > 0, m_1 + m_3 = 1$
In this case we have

$$e_2^T A m > e_3^T A m$$

contradicting $e_3 \in B(m)$.

3.) $m = (0, m_2, m_3), \;\; m_2, m_3 > 0, \;\; m_2 + m_3 = 1$
If (m, m) is a symmetric equilibrium, $e_2 \in B(m)$ and $e_3 \in B(m)$ must hold. This implies

$$e_2^T Am = e_3^T Am.$$

Taking into account $m_2+m_3 = 1$, this equation has a unique solution, namely $m_2 = \frac{1}{31}$, $m_3 = \frac{30}{31}$ with

$$e_2^T Am = e_3^T Am = \frac{36}{31}.$$

However,

$$e_1^T Am = \frac{45}{31} > \frac{36}{31}.$$

Thus, there is also no equilibrium where the support consists only of the second and the third strategy.

Finally, we have to show that no completely mixed symmetric equilibrium exists. The necessary condition

$$e_2^T Am = e_3^T Am$$

yields $m_3 = 30m_2$. However, since all $m_i > 0$, we have

$$e_1^T Am = 2m_1 + 45m_2 > 36m_2 = e_2^T Am = e_3^T Am.$$

This shows, that it is impossible to find a completely mixed vector $m \in \Delta^3$ with $Am = \lambda \mathbb{1}$ for a $\lambda \in \mathbb{R}$. Accordingly, there is no completely mixed symmetric equilibrium and (e_1, e_1) is unique.

B.3 The Games G1 and G2

The payoff matrix of G1 is given by (5.5). We observe that

$$B(e_1) = \{1\}$$
$$B(e_2) = \{1\}$$
$$B(e_3) = \{3\}.$$

Therefore, (e_1, e_1) and (e_3, e_3) are symmetric Nash equilibria. Obviously they are also both ESS. Note further that the second pure strategy is weakly dominated by e_1 and that

$$e_1^T Am > e_2^T Am$$

holds for any $m \neq e_3$. This implies that the second pure strategy can not occur in the support of any mixed symmetric equilibrium. For this reason, we only have a to consider mixed equilibria with $m = (m_1, 0, m_3),\ m_1, m_3 > 0$. The equation

$$e_1^T Am = e_3^T Am$$

together with $m_1 + m_3 = 1$ yields the unique solution $m = \frac{1}{7}(4, 0, 3)^T$.

All of these arguments apply also to G2 showing that these two games indeed have the same set of equilibria.

C. Proof of Proposition 6.3.1

Proposition 6.3.1. *Let P^b and P^s be finite sets of buyers' offers and sellers' asks. If there exist population distributions s^b and s^s such that (P^b, P^s, s^b, s^s) is an equilibrium then $P^b = P^s = P$, $m^b = m^s = m$ and $c \leq p_1, u \geq p_m$. On the other hand, if $P^b = P^s = P$, where P has only rational elements with $c \leq p_1, u \geq p_m$ and u and c are rational then there exists a population size n and population distributions s^b and s^s, such that (P^b, P^s, s^s, s^b) is an equilibrium.*

Proof. First the necessity of the given condition is established. It is trivial to see that $p_i^b \leq u$ and $p_i^s \geq c$ must hold in equilibrium. Assume that $P^b \neq P^s$. Then there either exist prices p_i^b, p_j^s, p_{j+1}^s with $p_i^b \in (p_j^s, p_{j+1}^s)$ or p_i^b, p_{i+1}^b, p_j^s with $p_j^s \in (p_i^b, p_{i+1}^b)$. Without restriction of generality let us assume that the former case holds. The expected utility of p_i^b is given by

$$U^b(p_i^b, P^s, s^s) = \sum_{k=1}^{j} s_k^s \Pi^b(p_i^b, p_k^s).$$

However, choosing the price p_j^s the buyer would expect the utility

$$U^b(p_j^s, P^s, s^s) = \sum_{k=1}^{j} s_k^s \Pi^b(p_j^s, p_k^s).$$

Since $p_i^b > p_j^s$ implies $\Pi^b(p_i^b, p) < \Pi^b(p_j^s, p)\ \forall p \leq p_j^s$ this contradicts the optimality of p_i^b. This establishes $P^s = P^b$ together with $c \leq p_1^b = p_1^s$, $u \geq p_m^s = p_m^b$ as a necessary equilibrium condition. To make notation easier, let us denote $P^b = P^s$ by P.

In order to show the second part of the claim, the equilibrium distributions s^b and s^s over P will be constructed. The arguments given above show that given the set P of asks any buyers' offer which is no element of P is necessarily suboptimal. Thus, in order to show that there exists a population distribution over sellers' prices such that all prices in P are optimal from a buyers' point of view, we have to find a population distribution s^s over P such that

$$U^b(p_i, P, s^s) = U^b(p_1, P, s^s) \quad \forall i = 1, \dots m. \tag{C.1}$$

Analogously we have to find an s^b such that

$$U^s(p_i, P, s^b) = U^s(p_1, P, s^b) \quad \forall i = 1, \dots m. \tag{C.2}$$

Let us first construct s^s. Note that

$$U^b(p_i, P, s^s) = \sum_{k=1}^{i} s_k^s \left(u - \frac{p_i + p_k}{2} \right)$$

which in particular implies

$$U^b(p_1, P, s^s) = s_1^s (u - p_1).$$

Thus, the system of equations (C.1) can be rewritten as

$$s_1^s (u - p_1) = \sum_{k=1}^{i} s_k^s \left(u - \frac{p_i + p_k}{2} \right) \; \forall i = 1, \dots, m. \tag{C.3}$$

Due to the triangular form of this system the solutions can be easily calculated as follows. We write $s_i^s = a_i s_1^s$, set $a_1 = 1$ and get from the i-th equation of system (C.3):

$$a_i = \frac{1}{u - p_i} \left(u - p_1 - \sum_{k=1}^{i-1} a_k \left(u - \frac{p_i + p_k}{2} \right) \right) \quad i = 2 \dots m.$$

Since a_{i-1} has been constructed such that

$$u - p_1 = \sum_{k=1}^{i-1} a_k \left(u - \frac{p_{i-1} + p_k}{2} \right) \quad \forall i = 2, \dots, m,$$

we get by using the inequality $p_{i-1} < p_i$ that $a_i > 0$ holds for all $i = 1, \dots, m$. Setting $s_1^s = (\sum_{i=1}^{m} a_i)^{-1}$ defines a probability distribution with support P and since all elements of P as well as u and c are rational all s_i^s are rational. In the same way a probability distribution s^b over P can be constructed such that (C.2) holds and all s_i^b are positive and rational. Accordingly, there exists an integer n such that all numbers s_i^s and s_i^b can be expressed by fractions with denominator n.

References

1. Adke, S.R. and Manjunath, S.M. (1984), *An Introduction to Finite Markov Processes*, Wiley.
2. Alander, J.T. (1998), An Indexed Bibliography of Genetic Algorithms in Economics, Report Series No. 94-1-ECO, University of Vasa.
3. Albin, P. and Foley, D.K. (1992), Decentralized dispersed exchange without an auctioneer, *Journal of Economic Behavior and Organization*, 18, 27-51.
4. Alchian, A.A. (1950), Uncertainty, Evolution and Economic Theory, *Journal of Political Economy*, 58, 211-222.
5. Anantharam, V. and Tsoucas, P. (1989), A Proof of the Markov Chain Tree Theorem, *Statistics & Probability Letters*, 8, 189-192.
6. Andreoni, J. and Miller, J.H. (1990), Auctions with Adaptive Artificially Intelligent Agents, Working Paper No. 91-01-004, Santa Fe Institute.
7. Arifovic, J. (1992), Genetic Algorithm: Experimental Evidence, Manuscript, McGill University, Montreal.
8. Arifovic, J. (1994), Genetic Algorithm learning and the Cobweb model, *Journal of Economic Dynamics and Control*, 18, 3-28.
9. Arifovic, J. (1995), Genetic Algorithms and Inflationary Economics, *Journal of Monetary Economics*, 36, 219-243.
10. Arifovic, J. (1996), The Behavior of Exchange Rate in the Genetic Algorithm and Experimental Economics, *Journal of Political Economy*, 104, 3, 510-541.
11. Arifovic, J. (1996), Strategic Uncertainty and the Genetic Algorithm Adaptation, Working Paper, Simon Fraser University, Vancouver.
12. Arifovic, J. and Eaton, C. (1998), The Evolution of Communication in a Sender/Receiver Game of Common Interest with Cheap Talk, *Journal of Economic Dynamics and Control*, 22, 8-9, 1187-1207.
13. Arthur, W.B., Holland, J.H., LeBaron, B., Palmer, R. and Tayler, P. (1997), Asset Pricing und Endogeneous Expectations in an Artificial Stock Market, in Arthur, W.B, Durlauf, S and Lane, D. (Eds.), *The Economy as an Evolving Complex System II*, Addison-Wesley, Reading, MA.
14. Ashlock, D., Smucker, M.D., Stanley, E.A. and Tesfatsion, L. (1996), Preferential Partner Selection in an Evolutionary Study of Prisoner's Dilemma, *BioSystems*, 37, 99-125.
15. Axelrod, R. (1984), *The Evolution of Cooperation*, Basic Books, New York.
16. Axelrod, R. (1987), The Evolution of Strategies in the Iterated Prisoner's Dilemma, in L. Davis (Ed.), *Genetic Algorithms and Simulated Annealing*, pp. 32-41, Pitman, London.
17. Azariadis, C. (1981), Self-Fulfilling Prophecies, *Journal of Economic Theory*, 25, 380-396.
18. Azariadis, C. and Guesnerie, R. (1986), Sunspots and Cycles, *Review of Economic Studies*, LIII, 725-737.

19. Beltrati, A. and Margarita, S. (1993), Evolution of Trading Strategies Among Heterogeneous Artificial Economic Agents, in J.A. Meyer, H.L. Roitblat and S.W. Wilson (Eds.), *From Animals to Animats 2, Proceedings of the Second International Conference on Simulation of Adaptive Behavior*, MIT Press, Cambridge, MA.
20. Benassy, J.P. and Blad, M.C. (1989), On learning and rational expectations in an overlapping generations model, *Journal of Economic Dynamics and Control*, 13, 379-400.
21. Berlekamp, E., Conway, J. and Guy, R. (1982), *Winning Ways*, vol. 2, Academic Press, New York.
22. Bernheim, B.D. (1984), Rationalizable strategic behavior, *Econometrica*, 52, 1007-1028.
23. Berninghaus, S. and Erhart, K.M. (1998), Long-run Evolution of Local Interaction Structures in Games, Working Paper, University of Karlsruhe.
24. Bezdek, J.C. (1992), On the Relationship Between Neural Networks, Pattern Recongnition and Intelligence, *International Journal of Approximate Reasoning*, 6, 85-107.
25. Biccheri, C. and Rovelli, C. (1995), Evolution and Revolution: The Dynamics of Corruption, *Rationality and Society*, 7, 201-224.
26. Biethan, J. and Nissen, V. (Eds.) (1995), *Evolutionary Algorithms in Management Applications*, Springer, Heidelberg.
27. Billingsley, P. (1968), *Convergence of Probability Measures*, Wiley, New York.
28. Blume, L.E. and Easley, D. (1982), Learning to Be Rational, *Journal of Economic Theory*, 26, 340-351.
29. Binmore, K. (1992), *Fun and Games*, D. C. Heath and Company, Lexington, Massachusetts.
30. Björnerstedt, J. and Weibull, J.W. (1996), Nash Equilibrium and Evolution by Imitation, in K. Arrow et al. (Eds.), *The Rational Foundations of Economic Behavior*, MacMillan, London, pp. 155-171.
31. Bray, M. (1982), Learning, Estimation and the Stability of Rational Expectations, *Journal of Economic Theory*, 26, 318-339.
32. Bray, M. and Kreps, D.M. (1987), Rational learning and Rational Expectations, in G.R. Feiwel (Ed.), *Arrow and the Ascent of Modern Economic Theory*, New York University Press, New York.
33. Bray, M.M. and Savin, N.E. (1986), Rational Expectations Equilibria, Learning, and Model Specification, *Econometrica*, 54, 1129-1160.
34. Brenner, T. (1996), Learning in a Repeated Decision Process: A Variation-Imitation-Decision Model, Papers on Economics and Evolution #9603, Max Planck Institut, Jena.
35. Brenner, T. (1998), Can Evolutionary Algorithms Describe Learning Processes?, *Journal of Evolutionary Economics*, 8, 271-284.
36. Bullard, J, and Duffy, J. (1998), A Model of Learning and Emulation with Artificial Adaptive Agents, *Journal of Economic Dynamics and Control*, 22, 179-207.
37. Carlson, J.A. (1968), An Invariably Stable Cobweb Model, *Review of Economic Studies*, 360-362.
38. Chaterjee, K. and Samuelson, W. (1983), Bargaining under Incomplete Information, *Operations Research*, 31, 835-851.
39. Conte, R., Hegselmann, R. and Terna, P. (Eds.) (1997), *Simulating Social Phenomena*, Springer, Heidelberg.
40. Cressman, R. (1992), *The Stability Concept of Evolutionary Game Theory*, Springer, Berlin.

41. Culberson, J.C. (1992), Genetic Invariance: A New Paradigm for Genetic Algorithm Design, Working Paper, University of Alberta.
42. Darwin, C. (1859), *On the Origin of Species by Means of Natural Selection*, John Murray, London.
43. Davis, T.E. and Principe, J.C. (1993), A Markov Chain Framework for the Simple Genetic Algorithm, *Evolutionary Computation*, 1.
44. Dawid, H. (1994), A Markov Chain Analysis of Genetic Algortihms with a State Dependent Fitness Function, *Complex Systems*, 8, 407-417.
45. Dawid, H. (1997), Learning of Equilibria by a Population with Minimal Information, *Journal of Economic Behavior and Organization*, 32, 1-18.
46. Dawid, H. (1996), Learning of Cycles and Sunspot Equilibria by Genetic Algorithms, *Journal of Evolutionary Economics*, 6, 361-373.
47. Dawid, H. (1998), On the Stability of Discrete Time Replicator Dynamics with Inertia, forthcoming in *Mathematical Social Sciences*.
48. Dawid, H. (1998), On the Convergence of Genetic Learning in a Double Auction Market, forthcoming in *Journal of Economic Dynamics and Control*.
49. Dawid, H. and Kopel, M. (1998), The Appropriate Design of a Genetic Algorithm in Economic Applications Exemplified by a Model of the Cobweb Type, *Journal of Evolutionary Economics*, 8, 297-315.
50. Dawid, H. and Mehlmann, A. (1996), Genetic Learning in Strategic Form Games, *Complexity*, 1, 5, 51-59.
51. Dawid, H. and Hornik, K. (1996), The Dynamics of Genetic Algorithms in Interactive Environments, *Journal of Network and Computer Applications*, 19, 5-19.
52. Day, R.H. (1975), Adaptive Processes and Economic Theory, in: R.H. Day and T. Groves (Eds.), *Adaptive Economic Models*, pp. 1-38, Academic Press.
53. Day, R.H. (1994), *Complex Economic Dynamics; Volume I: An Introduction to Dynamical Systems and Market Mechanisms*, MIT Press.
54. De Jong, K.A., Spears, W.M. and Gordon, D.F. (1995), Using Markov Chains to Analyze GAFOs, in L.D. Whitley and M.D. Vose (Eds.), *Foundations of Genetic Algorithms 3*, pp. 115-137, Morgan Kaufmann, San Mateo, CA.
55. Doob, J.L. (1953), *Stochastic Processes*, John Wiley & Sons, NY.
56. Dosi, G. and Nelson, R.R. (1994), An Introduction to Evolutionary Theories in Economics, *Journal of Evolutionary Economics*, 4, 153-172.
57. Eiben, A.E., Aarts, E.H.L. and Van Hee, K.M. (1991), Global Convergence of Genetic Algorithms: a Markov Chain Analysis, in H.-P. Schwefel and R. Männer (Eds.), *Parallel Problem Solving from Nature*, pp. 4-12, Springer.
58. Eichberger, J., Haller, H. and Milne, F. (1993), Naive Bayesian learning in 2×2 matrix games, *Journal of Economic Behavior and Organization*, 22, 69-90.
59. Epstein, J. and Axtell, R. (1996), *Growing Artificial Societies: Social Science form the Bottom Up*, MIT Press, Brookings, MA.
60. Fogel, D.B. (1993), Evolving Behaviors in the Iterated Prisoner's Dilemma, *Evolutionary Computation*, 1, 77-97.
61. Fogel, D.B. and Atmar, J.W. (1992), *Proceedings of the First Annual Conference on Evolutionary Programming*, Evolutionary Programming Society, San Diego, CA.
62. Fudenberg, D. and Levine, D.K. (1998), *The Theory of Learning in Games*, MIT Press, Cambridge, MA.
63. Fudenberg, D. and Tirole, J. (1991), *Game Theory*, MIT Press, Cambridge, MA.
64. Franke, R. (1998), Coevolution and Stable Adjustments in the Cobweb Model, to be published in *Journal of Evolutionary Economics*.

65. Friedman, D. (1991), Evolutionary Games in Economics, *Econometrica*, 59, 3, 637-666.
66. Freidlin, M.I. and Wentzell, A.D. (1984), *Random Perturbations of Dynamical Systems*, Springer.
67. Gardner, M. (1971), Mathematical Games, *Scientific American*, 224, February, 112.
68. Gardner, M. (1971), Mathematical Games, *Scientific American*, 224, March, 106.
69. Gardner, M. (1971), Mathematical Games, *Scientific American*, 224, April, 114.
70. Gardner, M. (1972), Mathematical Games, *Scientific American*, 226, January, 104.
71. Geyer-Schulz, A. (1992), On Learning in a Fuzzy Rule-Based Expert System, *Kybernetica*, 28, 33-36.
72. Goldberg, D.E. (1981), Robust Learning and Decision Algorithms for Pipeline Operations, unpublished dissertation proposal, University of Michigan, Ann Arbor.
73. Goldberg, D.E. (1987), Simple genetic algorithms and the minimal deceptive problem, in L. Davis (Ed.), *Genetic algorithms and simulated annealing*, pp. 32-41, Pitman, London.
74. Goldberg, D.E. (1989), *Genetic Algorithms in Search, Optimization and Machine Learning*, Addison-Wesley.
75. Goldberg, D.E. (1995), The Existential Pleasures of Genetic Algorithms, in G. Winter, J. Perioux, M. Galan and P. Cuesta (Eds.), *Genetic Algorithms in Engineering and Computer Science*, pp. 23-31, Wiley & Sons, Chichester.
76. Goldberg, D.E. and Bridges, C.L. (1990), An Analysis of a Reordering Operator on a GA-Hard Problem, *Biological Cybernetics*, 62, 397-405.
77. Goldberg, D.E. and Segrest, P. (1987), Finite Markov Chain Analysis of Genetic Algorithms, in J.J. Grefenstette (Ed.), *Genetic Algorithms and their Applications: Proceedings of the Second International Conference on Genetic Algorithms*, pp. 1-8, Lawrence Earlbaum Associates, Hilsdale, New Jersey.
78. Hanne, T. (1994), Prisoner's Dilemmas Can Survive Evolutionary Algorithms, Working Paper, Fern Universität Hagen.
79. Harsanyi, J.C. and Selten, R. (1988), *A General Theory of Equilibrium Selection in Games*, MIT Press, Cambridge.
80. Hartl, R.F. (1991), A Global Convergence Proof for a Class of Genetic Algorithms, unpublished manuscript, Vienna University of Technology.
81. Hetz, J., Krogh, A. and Palmer, R. (1991), *Introduction to the Theory of Neural Computation*, Addison-Wesley, Redwood City, CA.
82. Hill, T.L. (1966), Studies in irreversible thermodynamics IV. Diagrammatic representation of steady state fluxes for unimolecular systems, *Journal of Theoretical Biology*, 10, 442-459.
83. Hinterding, R., Michalewicz, Z. and Peachey T.C. (1996), Self-Adaptive Genetic Algorithm for Numeric Functions, in H.-M. Voigt, W. Ebeling, I. Rechenberg and H.-P. Schwefel (Eds.), *Parallel Problem Solving from Nature*, pp. 420-429, Springer, Berlin.
84. Ho, T.-H. (1996), Finite Automata Play Repeated Prisoner's Dilemma with Information Processing Costs, *Journal of Economic Dynamics and Control*, 20, 173-207.
85. Hofbauer, J. and Sigmund, K. (1988), *The Theory of Evolution and Dynamical Systems*, Cambridge University Press, Cambridge.
86. Hofbauer, J. and Sigmund, K. (1998), *Evolutionary Games and Population Dynamics*, Cambridge University Press, Cambridge.

87. Holland, J.H. (1975), *Adaptation in Natural and Artificial Systems*, The University of Michigan Press, Ann Arbor, MI.
88. Holland, J.H. (1992), *Adaptation in Natural and Artificial Systems: An Introductory Analysis with Applications to Biology, Control, and Artificial Intellingence*, MIT Press, Cambridge, MA.
89. Holland, J.H. and Miller, J.H. (1991), Artificial adaptive agents in economic theory, *American Economic Review: Papers and Proceedings of the 103rd Annual Meeting of the American Economic Association*, 335-350.
90. Hornik, K., Stinchcombe, M. and White, H. (1989), Multilayer Feedforward Networks are Universal Approximators, *Neural Networks*, 2, 359-366.
91. Kandori, M., Mailath, G.J. and Rob, R. (1993), Learning, Mutation, and Long Run Equilibria in Games, *Econometrica*, 61, 29-56.
92. Kandori, M. and Rob, R. (1995), Evolution of Equilibria in the Long Run: A General Theory and Applications, *Journal of Economic Theory*, 65, 383-414.
93. Keenan, D.C. and O'Brien, M.J. (1993), Competition, Collusion, and Chaos, *Journal of Economic Dynamics and Control*, 17, 327-353.
94. Kelley, W.G. and Peterson, A.C. (1991), *Difference Equations, An Introduction with Applications*, Academic Press.
95. Keerthi, S.S. and Ravindran, B. (1994), A Tutorial Survey of Reinforcement Learning, Working Paper, Department of Computer Science and Automation, Indian Institute of Science.
96. Kirchkamp, O. (1996), Spatial Evolution of Automata in the Prisoners' Dilemma — Preliminary Results, in K. G. Troitzsch, U. Mueller, G. N. Gilbert, J. E. Doran (Eds.) *Social Science Microsimulation*, Springer, Heidelberg.
97. Kirman, A. and Salmon, M. (Eds.) (1995), *Learning and Rationality in Economics*, Blackwell, Oxford.
98. Kirman, A. and Vriend, N.J. (1997), Evolving Market Structure: A Model of Price Dispersion and Loyalty, Working Paper, University of London.
99. Kiyotaki, N. and Wright, R. (1989), On Money as a Medium of Exchange, *Journal of Political Economy*, 97, 927-954.
100. Koehler, G.J. (1994), A Proof of the Vose-Liepins Conjecture, *Annals of Mathematics and Artificial Intelligence*, 5, 79-88.
101. Koehler, G.J., Bhattacharyya, S. and Vose, M.D. (1997), General Cardinality Genetic Algorithms, *Evolutionary Computation*, 5, 4, 439-459.
102. Kopel, M. (1994), *Komplexe Unternehmensdynamik, Chaotische Dynamische Systeme in der Betriebswirtschaftslehre*, Deutscher Universitätsverlag.
103. Koza, J.R. (1991), *Genetic Programming*, MIT Press, Cambridge, MA.
104. Krishna, V. (1992), Learning in Games with Strategic Complementarities, mimeo, Harvard Business School
105. Laffond, G. and Lesourne, J. (1992), The genesis of expectations and of sunspot equilibria, *Journal of Evolutionary Economics*, 2, 211-231
106. Leighton, F.T. and Rivest, R.L. (1983), The Markov Chain Tree Theorem, M.I.T. Laboratory for Computer Science, Technical Report, MIT/LCS/TM-249.
107. Leighton, F.T. and Rivest, R.L. (1986), Estimating a probability using finite memory, *IEEE Transactions on Information Theory*, 32, 733-742.
108. Leontief, W.W. (1934), Verzögerte Angebotsanpassung und partielles Gleichgewicht, *Zeitschrift für Nationalökonomie*, 5, 670-676.
109. Lettau, M. (1997), Explaining the Facts with Adaptive Agents: The Case of Mutual Funds, *Journal of Economic Dynamics and Control*, 21, 1117-1147.
110. Litterman, R. and Knez, P. (1989), Genetic Algorithm for the Kiyotaki-Wright Model, Mimeo, Goldman-Sachs.

111. Lucas, R.E. (1986), Adaptive Behavior and Economic Theory, *Journal of Business*, 59, 401-426.
112. Mailath, G.J. (1992), Introduction: Symposium on Evolutionary Game Theory, *Journal of Economic Theory*, 57, 259-277.
113. Marimon, R., McGrattan, E. and Sargent, T. (1990), Money as a Medium of Exchange in an Economy with Artificially Intelligent Agents, *Journal of Economic Dynamics and Control*, 14, 329-374.
114. Marcet, A. and Sargent, T.J. (1989), Convergence of Least Squares Learning Mechanisms in Self Referential Linear Stochastic Models, *Journal of Economic Theory*, 48, 337-368.
115. Marcet, A. and Sargent, T.J. (1989), Convergence of Least Squares Learning in Environments with Hidden State Variables and Private Information, *Journal of Political Economy*, 97, 1306-1322.
116. Marimon, R. and Sunder, S. (1993), Indeterminacy of Equilibria in a Hperinflationary Wrld: Experimental Evidence, *Econometrica*, 61, 1073-1108.
117. Marks II, R.J. (1993), Intelligence: Computational Versus Artificial, *IEEE Transactions on Neural Networks*, 4, 737-739.
118. May, R.M. and Nowak, M.A. (1992), Evolutionary Games and Spatial Chaos, *Nature*, 359, 826-829.
119. May, R.M. and Nowak, M.A. (1993), The Spatial Dilemma of Evolution, *Journal of Bifurcation and Chaos*, 3, 35-78.
120. Michalewicz, Z. (1994), *Genetic Algorithms + Data Structures = Evolution Programs*, Springer.
121. Midgley, D.F., Marks, R.E. and Cooper, L.G. (1997), 'Breeding Competitive Strategies', *Management Science*, 43, 257-275.
122. Mitchell, M. (1995), *An Introduction to Genetic Algorithms*, MIT Press, Cambridge, MA.
123. Miyazawa, K. (1961), On the Convergence of the Learning Process in a 2×2 Non-zero-sum Game, Econometric Research Program, Research Memorandum No. 33, Princeton University, Princeton.
124. Moss, S. and Rae, J. (Eds.) (1992), *Artificial Intelligence and Economic Analysis*, Edward Elgar.
125. Muth, J.F. (1961), Rational Expectations and the Theory of Price Movements, *Econometrica*, 29, 315-335.
126. Mühlenbein, H. (1991), Darwin's Continent Cycle Theory and Its Simulation by the Prisoner's Dilemma, *Complex Systems*, 5, 459-478.
127. Mühlenbein, H. (1991), Evolution in Time and Space- the Parallel Genetic Algorithm, in G. Rawlins (Ed.), *Foundations of Genetic Algorithms*, pp. 316-337, San Mateo: Morgan Kaufman.
128. Mühlenbein, H. (1997), Genetic Algorithms, in E.H.L. Aarts and J.K. Lenstra (Eds.), *Local search in Combinatorial Optimization*, pp. 137-171, Wiley, Chichester.
129. Mühlenbein, H. (1997), The Equation for Response to Selection and its Use for Prediction, *Evolutionary Computation*, 5, 303-346.
130. Mühlenbein, H. and Schlierkamp-Voosen (1994), The science of breeding and its application to the breeder genetic algorithm BGA, *Evolutionary Computation*, 1, 4, 335-360.
131. Nash, J.F. (1951), Non-cooperative Games, *Annals of Mathematics*, 54, 286-295.
132. Nerlove, M. (1958), Adaptive Expectations and Cobweb Phenomena, *Quarterly Journal of Economics*, 72, 227-240.
133. Nix, A.E. and Vose, M.D. (1992), Modeling Genetic Algorithms with Markov Chains, *Annals of Mathematics and Artificial Intelligence*, 5, 79-88.

134. Novkovic, S. and Sverko, D. (1997), Modeling Heterogeneous Agents with Induced Population Diversity. Proceedings of the Workshop on Economics with Heterogeneous Interacting Agents, Ancona.
135. Nyarko, Y., Woodford, M. and Yannelis, N.C. (1994), Bounded Rationality and learning: Introduction, *Economic Theory*, 4, 811-820.
136. Pearce, D, (1984), Rationalizable strategic behavior and the problem of perfection, *Econometrica*, 52, 1029-1050.
137. Pingle, M. (1995), Imitation versus Rationality: An Experimental Perspective in Decision Making, *Journal of Socio-Economics*, 24, 281-315.
138. Radner, R. and Schotter, A.(1989), The Sealed-Bid Mechanism: An Experimental Study, *Journal of Economic Theory*, 48, 179-220.
139. Rechenberg, I. (1994), *Evolutionsstrategie '94*, Frommann-Holzboog, Stuttgart-Bad Cannstadt.
140. Rehkugler, H. and Zimmermann, H.G. (1994), *Neuronale Netze in der Ökonomie: Grundlagen und finanzwirtschaftliche Anwendungen*, Vahlen, München.
141. Robbins, J. and Monro, S. (1951), A Stochastic Approximation Method, *Annals of Mathematical Statistics*, 22, 400-407.
142. Robinson, J. (1951), An Iterative Method of Solving a Game, *Annals of Mathematics*, 54, 296-301.
143. Robson, A. (1993), Random Pairing and Fast Convergence to Efficient Equilibria in Coordination Games, Working Paper, University of Western Ontario.
144. Rudolph, G. (1994), Convergence Analysis of Canonical Genetic Algorithms, *IEEE Transactions on Neural Networks*, 5, 96-101.
145. Rust, J., Palmer, R. and Miller, J.H. (1992), Behavior of Trading Automata in a Computerized Double Auction Market, in D. Friedman and J. Rust (Eds.), *The double auction market: Institutions, theories and evidence*, Addison-Wesley, Redwood City, CA.
146. Rust, J., Miller, J.H. and Palmer, R. (1994), Characterizing Effective Trading Strategies: Insights from a Computerized Double Auction Tournament, *Journal of Economic Dynamics and Control*, 18, 61-96.
147. Samuelson, L. (1997), *Evolutionary Games and Equilibrium Selection*, MIT Press, Cambridge, MA.
148. Samuelson, P. (1958), An Exact Consumption-Loan Model of Interest with or without the Social Contrivance of Money, *Journal of Political Economy*, 66, 467-482.
149. Samuelson, L. (1994), Stochastic Stability in Games with Alternative Best Replies, *Journal of Economic Theory*, 64, 35-65.
150. Sargent, T.J. (1993), *Bounded Rationality in Macroeconomics*, Clarendon Press, Oxford.
151. Schaffer, J.D., Caruna, R.A., Eshelman L.J. and Das, R. (1989), A study of control parameters affecting online performance of genetic algorithms for function optimization, in J.D. Schaffer (Ed.), *Proceedings of the Third International Conference on Genetic Algorithms*, pp. 51-60, Morgan Kaufmann.
152. Schaffer, J.D. and Eshelman, L.J. (1991), On crossover as an evolutionarily viable strategy, in R.K. Belew and L.B. Booker (Eds.), *Proceedings of the Fourth International Conference on Genetic Algorithms*, pp. 61-68, Morgan Kaufmann.
153. Schwefel, H.P. (1977), *Numerische Optimierung von Computer-Modellen mittels der Evolutionsstrategie*, Birkhäuser, Basel.
154. Selten, R. (1975), Reexamination of the Perfectness Concept for Equilibrium Points in Extensive Games, *International Journal of Game Theory*, 4, 25-55.

155. Selten, R. (1980), A Note on Evolutionarily Stable Strategies in Asymetrical Animal Conflicts, *Journal of Theoretical Biology*, 84, 93-101.
156. Selten, R. (1991), Evolution, Learning, and Economic Behavior, *Games and Economic Behavior*, 3, 3-24.
157. Shapley, L. (1964), Some Topics in Two-Person Games, in M. Dresher et al. (Eds.), *Advances in Game Theory, Annals of Mathematical Studies*, 52, 1-28.
158. Shubert, B.O. (1975), A flow graph formula for the stationary distribution of a Markov chain, *IEEE Transactions on Systems, Man and Cybernetics*, 5, 565-566.
159. Simon, H.A. (1982), *Models of Bounded Rationality*, MIT Press, Cambridge.
160. Sims, C.A. (1980), Macroeconomics and Reality, *Econometrica*, 69, 1-48.
161. Smith, R.E. and Brown Cribbs III, H. (1994), Is a Learning Classifier System a Type of Neural Network?, *Evolutionary Computation*, 2.
162. Spears, W.M. (1995), Adapting Crossover in Evolutionary Algorithms, in J.R. McDonnell, R.G. Reynolds and D.B. Fogel (Eds.) *Proceedings of the Fourth Annual Conference on Evolutionary Programming*, pp. 367-384, MIT Press, Cambridge.
163. Suzuki, J. (1993), A Markov Chain Analysis on a Genetic Algorithm, in S. Forrest (Ed.), *Proceedings of the Fifth International Conference on Genetic Algorithms*, pp. 146-153, Morgan Kaufmann.
164. Suzuki, J. (1995), A Markov Chain Analysis on Simple Genetic Algorithms, *IEEE Transactions on Systems, Man and Cybernetics*, 25, 655-659.
165. Srinivas, M. and Patnaik, L.M. (1994), Adaptive Probabilities of Crossover and Mutation in Genetic Algorithms, *IEEE Transactions on Systems, Man, and Cybernetics*, 24, 17-26.
166. Tesfatsion, L. (1997), A Trade Network Game with Endogeneous Partner Selection, in Amman, H., Rustem, B. and Whinston, A.B. (Eds.), *Computational Approaches to Economic Problems*, pp. 249-269, Kluwer.
167. Turnovsky, S.J. (1969), A Bayesian Approach to the Theory of Expectations, *Journal of Economic Theory*, 1, 220-227.
168. Valley, K., Thompson, L., Gibbons, R. and Bazerman, M.H. (1995), Is Talk Really Cheap? Outperforming Equilibrium Models of Communication in Bargaining Games, Working Paper, Harvard University.
169. Van Damme, E. (1987), *Stability and Perfection of Nash Equilibria*, Springer, Berlin.
170. Van Huyck, J., Battalio, R. and Beil, R. (1990), Tacit Coordination Games, Strategic Uncertainty, and Coordination Failure, *American Economic Revue*, 80, 234-248.
171. Vega-Redondo, F. (1995), Expectations, Drift, and Volatility in Evolutionary Games, *Games and Economic Behavior*, 11, 391-412.
172. Vega-Redondo, F. (1996), *Evolution, Games and Economic Behavior*, Oxford University Press, Oxford.
173. Vega-Redondo, F. (1997), The Evolution of Walrasian Behavior, *Econometrica*, 65, 375-384.
174. Von Neumann, J. (1966), *Theory of Self-Reproducing Automata*, edited by A.W. Burks, University of Illinois, Urbana,IL.
175. Vose, M.D. and Liepins, G.E. (1991), Punctuated Equilibria in Genetic Search, *Complex Systems*, 5, 31-44.
176. Vose, M.D. and Wright, A.H. (1994), Simple Genetic Algorithms with Linear Fitness, *Evolutionary Computation*, 2, 4, 347-368.
177. Vose, M.D. and Wright, A.H. (1996), Stability of Vertex Fixed Points and Applications, in L.D. Whitley and M.D. Vose (Eds.), *Foundations of Genetic Algorithms 3*, pp. 103-113, Morgan Kaufmann, San Mateo, CA.

178. Vriend, N.J. (1995), Self-Organization of Markets: An Example of a Computational Approach, *Computational Economics*, 8, 205-231.
179. Vriend, N.J. (1998), An Illustration of the Essential Difference between Individual and Social Learning, and its Consequences for Computational Analyses, Working Paper, University of London.
180. Weibull, J.W. (1995), *Evolutionary Game Theory*, MIT Press, Cambridge, MA.
181. Weissing, F.J. (1991), Evolutionary Stability and Dynamic Stability in a Class of Evolutionary Normal Form Games, in R. Selten (Ed.), *Game Equilibrium Models I*, Springer.
182. Welford, C.P. (1989), A Laboratory Analysis of Price Dynamics and Expectations in the Cobweb Model', Discussion Paper 89-15, University of Arizona.
183. White, H. (1992), *Artificial Neural Networks: Approximation and Learning Theory*, Basil Blackwell, Oxford.
184. Whitley, D. (1989), The GENITOR algorithm and selection pressure: why rank-based allocation of reproductive trials is best, in J.D. Schaffer (Ed.) *Proceedings of the Third International Conference on Genetic Algorithms*, pp. 116-121, Morgan Kaufmann.
185. Wolfram, S. (1994), *Cellular Automata and Complexity: Collected Papers*, Addison-Wesley, Reading, Ma.
186. Woodford, M. (1990), Learning to Believe in Sunspots, *Econometrica*, 58, 277-307.
187. Wright, A.H. and Vose, M.D., Finiteness of the Fixed Point Set for the Simple Genetic Algorithm, *Evolutionary Computation*, 3, 3, 299-309.
188. Wuensche, A. and Lesser, M. (1992), *The Global Dynamics of Cellular Automata*, Addison-Wesley, Reading, Ma.

List of Figures

2.1 A simple example of a message and a rule list in a classifier system; the first and the fourth rule are matched 15
2.2 The structure of a classifier system 16
2.3 Example of a threshold and a sigmoid logistic activation function . 18
2.4 A feedforward network with one hidden layer (a 3-5-2 network)... 19
2.5 Example of a local rule of a cellular automaton. According to this rule the value of the central cell is the exclusive or of the previous values of the direct neighbors. The short name of this rule is rule 90 or modulo 2 rule 22
2.6 The Von Neumann and Moore neighborhood of a site. The cells with bold faced values are in the neighborhood of the central cell. The left picture shows the Von Neumann neighborhood, the right one the Moore neighborhood 23
2.7 The evolution of a rule 90 automaton starting with an initial condition with only one cell with value 1. As well the global effects of the different value of this single cell as the self organizing behavior can be seen quite clearly 25
2.8 An example of the Axelrod like encoding of a strategy in the IPD. The action part of the string contains 64 bits 29

3.1 The structure of a Genetic Algorithm 42
3.2 One-point crossover with crossover point 2 45
3.3 Uniform crossover with a given crossover mask 46
3.4 The fitness function to be maximized by the genetic algorithm ... 47

4.1 The average population strategy in the iterated prisoner's dilemma according to the approximating deterministic dynamics and in a real simulation with a simple GA ($n = 2000,\ l = 6,\ \chi = 1,\ \mu = 0.001$) 84
4.2 The fraction of individuals playing the pure strategy "defect" according to the approximating deterministic dynamics and in a real simulation with a simple GA ($n = 2000,\ l = 6,\ \chi = 1,\ \mu = 0.001$) 84

5.1 The average population strategy of a SGA in a RSP game with the parameter values $(x, y, z) = (1.1, 4, 1)$, $n = 100$, $\chi = 0.8$ and $\mu = 0.001$ 104
5.2 The average population strategy of a SGA in a RSP game with the parameter values $(x, y, z) = (1.1, 4, 1)$, $n = 100$, $\chi = 0.8$ and $\mu = 0$ 104
5.3 The average population strategy of a SGA in a RSP game with the parameter values $(x, y, z) = (2.4, 4, 1)$, $n = 100$, $\chi = 0.8$ and $\mu = 0.001$ 106
5.4 The average population strategy of a SGA in a RSP game with the parameter values $(x, y, z) = (3.5, 4, 1)$, $n = 100$, $\chi = 0.8$ and $\mu = 0.001$ 106
5.5 The average population strategy of a SGA in the GA deceptive game GAD with the parameter values $n = 100$, $\chi = 0.8$ and $\mu = 0$ 109
5.6 The average population strategy of a SGA in the GA deceptive game GAD with the parameter values $n = 100$, $\chi = 0.8$ and $\mu = 0.001$ 109
5.7 The average population strategy of a SGA in the GA deceptive game GAD with the parameter values $n = 100$, $\chi = 0.8$, $\mu = 0$ and one conservative agent for each pure strategy 110
5.8 The average population strategy of a SGA in the GA deceptive game GAD with the parameter values $n = 100$, $\chi = 0.8$, $\mu = 0$ and three conservative agents for each pure strategy 110
5.9 The average population strategy converging towards the risk dominant equilibrium in $G1$ ($n = 100$, $\chi = 0.8$, $\mu = 0$) 113
5.10 The average population strategy converging towards the risk dominated equilibrium in $G2$ ($n = 100$, $\chi = 0.8$, $\mu = 0$) 113
5.11 The trajectories of the average population strategies in a two population SGA in a RSP game with parameters $(x, y, z) = (1.1, 4, 1), n^1 = n^2 = 100, \chi^1 = \chi^2 = 0.8, \mu^1 = \mu^2 = 0.001$ 116
5.12 The trajectories of the average population strategies in a two-population SGA in the GA deceptive game GAD with parameters $n^1 = n^2 = 100, \chi^1 = \chi^2 = 0.8, \mu^1 = \mu^2 = 0$ 117
5.13 The trajectories of the average population strategies in a two-population SGA in the GA deceptive game GAD with parameters $n^1 = n^2 = 100, \chi^1 = \chi^2 = 0.8, \mu^1 = \mu^2 = 0.001$ 117
5.14 The trajectories of the average population strategies in a two-population SGA in the GA deceptive game GAD with parameters $n^1 = n^2 = 100, \chi^1 = \chi^2 = 0.8, \mu^1 = \mu^2 = 0$ and one conservative agent for each pure strategy in both populations 119

6.1 The average production quantity in P_t for $a = 5,\ b = 5,\ \alpha = 0.25$ and $\beta = 1$ 125
6.2 The average production quantity in P_t for $a = 5,\ b = 5,\ \alpha = 1$ and $\beta = 1$ 125
6.3 The average production quantity in P_t for $a = 5,\ b = 5,\ \alpha = 1,\ \beta = 1$ and one special bit 128
6.4 The average production quantity in P_t for $a = 5,\ b = 5,\ \alpha = 1,\ \beta = 1$ and two adjacent special bits 128
6.5 The average production quantity in P_t for $a = 5,\ b = 5,\ \alpha = 1,\ \beta = 1$ and two separated special bits 131
6.6 The utility function $U(c_1, c_2)$ 137
6.7 The evolution of the population averages of the expected real return of money in both populations ($n = 30,\ l = 20,\ \chi = 0.6$ and $\mu = 0.001$) 139
6.8 The observed real return of money for the same simulation as in figure 6.7 139
6.9 The evolution of the population averages of the expected value of $\frac{p0}{p1}$, when both populations believe in sunspots ($n = 30,\ l = 20,\ \chi = 0.6$ and $\mu = 0.001$) 142
6.10 The observed real return of money for the same simulation as in figure 6.9 142
6.11 The empirical distribution of trading prices in a double auction market. 151
6.12 The empirical distribution of the difference between the long run bids of buyers and sellers. 151
6.13 The average population values of offers and asks in a typical simulation run. 152
6.14 The empirical distribution of long run trading prices with $c = 0.1$ and $u = 0.7$. 154
6.15 The empirical distribution of long run trading prices with uniform initializations. 154
6.16 The long run outcomes of the buyers' and sellers' strategy parameters in 50 different simulation runs with stochastic cost and utility values 158
6.17 Buyers' (a) and sellers' (b) expected payoff of an interaction with an opponent using the equilibrium strategy in dependence of the own strategy parameters. 159

7.1 A graphic illustration of the stability criterion for the rational expectations equilibrium in the cobweb model 163
7.2 Graphic stability criterion for the rational expectations equilibrium in the cobweb model 163
7.3 Graphical stability check for the uniform population with average production of $\bar{y} = 0.75$ 164

7.4 Graphical stability check for the uniform population with average production of $\bar{y} = 0.25$ 164
7.5 Graphical stability check for the "false" homogeneous equilibrium state in the case of pure quantity decisions........ 166
7.6 Graphical stability check for the "false" homogeneous equilibrium state in the case of separated exit and entry decisions 166

List of Tables

6.1 The results of the GA in a model with stationary and cyclical equilibria with $n = 30$, $l = 20$, $\chi = 0.6$ and $\mu = 0.1$ 138
6.2 The results of the GA incorporating sunspot depending expectations with $n = 30$, $l = 20$, $\chi = 0.6$ and $\mu = 0.1$. 141

Index

accumulation point, 59
adaptation, 9
advertising, 32
agents
– artificially intelligent, 12
– conservative, 108, 118
artificial intelligence, 1
auction, 26
– sealed bid double, 143
Axelrod's tournament, 28

back-propagation, 20
Bayesian learning, 99
best reply, 26
bucket brigade algorithm, 15
building block hypothesis, 46, 51

canonical genetic algorithm, 56
cartel building, 36
cellular automaton, 21–25, 31, 32, 35
– elementary, 21
– legal, 22
– totalistic, 22
Chebyshev's inequality, 79
classifier system, 14–17, 33, 35, 36, 41
co-evolution, 71
cobweb model, 32, 121–124, 161
coding mechanism, 90
communication effect, 67, 88
comparative static, 8
computational intelligence, 1
continent cycle theory, 29
convergence time, 56
cooperation, 27
crossover, 45–46, 51, 67
– multi-point, 45
– one-point, 45, 48, 51, 87, 90
– uniform, 45, 53, 54, 88

defection, 27
defining length, 50, 51
don't care symbol, 14, 50
duopoly, 7, 16
dynamical system, 77, 173–174

econometrics, 10
election, 46
election operator, 32
elitist strategy, 57
empirical evidence, 61
equilibrium, 1
– economic, 71, 89
– Nash, 26, 98, 114
– one-price, 145
– periodic, 134
– rational expectations, 133, 161
–– heterogeneous, 129
–– homogeneous, 123
– strict economic, 71, 89
– sunspot, 135
evolution
– natural, 61
evolutionary programming, 13
evolutionary stability, 98, 114
evolutionary strategies, 13

fiat money, 133
fictitious play, 3, 99
fitness function, 13, 42
– one-max, 53–54
– state dependent, 72
fitness scaling, 44
fixed costs, 122

GA deceptive
– function, 108
– game, 107, 116
game theory
– evolutionary, 4, 27, 97–101
genetic drift, 56
genetic operator, 13, 43–46
genetic programming, 13
genetic waste, 90, 165
graph, 74

heritability, 52

imitation effect, 67
implicit parallelism, 50
information exchange, 88
innovation, 68
inverse demand function, 122
inversion, 46

laboratory experiments, 61
learning, 9
– Bayesian, 10
– least square, 10
– moving average, 10
Q-, 20
– reinforcement, 20
– supervised, 20
– TD(λ), 20
– unsupervised, 20
limit distribution, 73, 172

Markov chain tree theorem, 74, 172
Markov theory, 2, 55, 73, 171–173
mating pool, 43
message list, 14
mixing matrix, 59, 82
mixing operator, 60
money model of, 36
Moore machine, 30
mutation, 41, 46, 48, 51, 68

neighborhood structure, 21
neural network, 17–21, 34
neuron, 17

oligopoly, 67
order, 50
overlapping generations model, 37, 132

perfect foresight, 123, 133
population
– initial, 42
– size, 42
– uniform, 43, 76
premature convergence, 43
prisoner's dilemma, 27, 82
– iterated, 28, 68

quantitative genetics, 52

rational expectations, 7–10
rationality
– bounded, 7, 9
– complete, 7
reordering, 46
replicator dynamics, 78, 89, 101
representative individual, 11
response to selection, 52
risk dominance, 100, 112
rock-scissors-paper game, 101, 103, 115
rule list, 14

satisficing, 10
schema, 50
schema theorem, 48, 50–52
selection, 43–44, 51, 67
– binary, 44
– fitness rank, 44
– natural, 41
– proportional, 43, 47, 50, 59
– tournament, 44
– truncation, 44, 54
selection differential, 52
selection intensity, 52, 54
self organization, 34, 36
simple genetic algorithm, 42, 47, 82
slow finishing, 44
stability, 174
– of heterogeneous states, 76
– of uniform states, 83–90
stationary distribution, 56, 59, 171
stochastic approximation, 10
stock market, 33, 34
stopping condition, 43
strategy
– dominance of, 28
– elitist, 49
– linear bidding, 156
stylized facts, 62
sunspots, 134

Tit-for-Tat, 28
transition matrix, 58, 171

VID model, 66

Druck: Strauss Offsetdruck, Mörlenbach
Verarbeitung: Schäffer, Grünstadt